纺织服装高等教育"十三五"部委级规划教材

服装英语
（英汉对照）
服装专业入门

高秀明　编著

东华大学出版社
·上海·

本书由扬州大学出版基金资助

图书在版编目（CIP）数据

服装英语 / 高秀明编著 . —上海：东华大学出版社，2018.12

ISBN 978-7-5669-1488-0

Ⅰ.①服… Ⅱ.①高… Ⅲ.①服装工业—英语 Ⅳ.① TS941

中国版本图书馆 CIP 数据核字（2018）第 242007 号

责任编辑：徐建红
封面设计：贝　塔

服装英语（英汉对照）
FUZHANG YINGYU

高秀明　编著

出　　　版：东华大学出版社（上海市延安西路1882号，200051）
本 社 网 址：dhupress.dhu.edu.cn
天猫旗舰店：http://dhdx.tmall.com
营 销 中 心：021-62193056　62373056　62379558
印　　　刷：苏州望电印刷有限公司
开　　　本：787 mm × 1092 mm　1/16
印　　　张：11.5
字　　　数：330 千字
版　　　次：2018 年 12 月第 1 版
印　　　次：2018 年 12 月第 1 次印刷
书　　　号：ISBN 978-7-5669-1488-0
定　　　价：49.80 元

Contents

Part One

Passage 1 Clothing ·· 2

Passage 2 Fashion ·· 12

Passage 3 The Types of Fashion ························ 24

Passage 4 Fashion Design ··································· 36

Passage 5 Color ·· 48

Passage 6 Fashion Drawing ································ 56

Passage 7 Fashion Show ···································· 66

Part Two

Passage 8   Pattern Cutting ···································· 80

Passage 9   Draping ·· 90

Passage 10 Pants Pattern ···································· 104

Passage 11 Bodice Pattern ·································· 118

Passage 12 Sleeve Pattern ·································· 130

Passage 13 Basic Pattern Techniques of Jackets ···· 142

Passage 14 Basic Sewing Techniques of Jackets ···· 154

Glossary ··· 172

References ··· 178

Postscript ·· 179

目　录

第一部分

- 第 1 课　服装 ··· 3
- 第 2 课　时尚 ··· 13
- 第 3 课　时装类型 ··· 25
- 第 4 课　时装设计 ··· 37
- 第 5 课　色彩 ··· 49
- 第 6 课　时装画 ·· 57
- 第 7 课　时装秀 ·· 67

第二部分

- 第 8 课　服装纸样裁剪 ·· 81
- 第 9 课　服装立体裁剪 ·· 91
- 第 10 课　裤子纸样 ·· 105
- 第 11 课　衣身原型纸样 ·· 119
- 第 12 课　袖子纸样 ·· 131
- 第 13 课　上装基础纸样技术 ··· 143
- 第 14 课　上装基础缝制工艺 ··· 155

词汇表 ··· 172

参考资料 ·· 178

后记 ·· 179

1

Part One

第一部分

Passage 1

Clothing

Clothing[1] is fiber[2] and textile[3] material worn on the body. The wearing[4] of clothing is mostly restricted to human beings and is a feature of nearly all human societies. The amount and type of clothing worn depends on physical, social and geographic considerations. Some clothing types can be gender-specific, although this does not apply to cross dressers[5].

1. Clothing, body and fashion[6]

No reason exists for feeling that one's body is forbidden topic. What might make it seem so are the associations of fear it has come to carry. After all, it is only in very recent times that the body, after being assiduously① concealed for hundreds of years, has been revealed. This process, once begun, seems almost to be escalating. The increasing manifestation of the body has caused a measure of surprise, and the phenomenon is still regarded with deep suspicion②.

It must, nevertheless, be acknowledged that we relate many of our concepts of nature to the human body. We refer, for example, to "the foot" of a mountain and "the neck" of a river isthmus③, and we say that a tree "stands". Michelangelo extended this parallel when he stated, "The man who cannot master the human body, and particularly its anatomy, will never understand the meaning of architecture."

All our ideas of proportion[7] are related to the body. Our movements arise out of our sight and our instinctive muscular reactions. Whether we are sleeping or walking, our feeling for rhythm[8] is closely connected with our regular heartbeats and the rise and fall of our breathing.

Clothes assume significance only when they are on the body. When they are hung up in a wardrobe[9] they look pathetically④ helpless; they seem to be voicelessly denouncing the cruelty of the tailor who forced them into their state of sad dependence. To really understand clothes it is necessary first to see the reasoning behind them and then to see them, as it were, in action. Clothes are more than just products of a textile factory or exhibits in a museum; they are artefacts⑤, used by people in all activities of daily life—standing, sitting, dancing, working or dying. Their true significance only becomes apparent[10] when we consider how they are related and adapted to the body. So many different human types exist: thin people, fat people, people with large heads, pin-headed people. But human nature being perverse⑥, styles do not always echo the body framework. If the body does not suit[11] a certain style[12] of dress then it is the clothes and not the body which should be modified.

第 1 课

服　装

　　服装是穿在人身体上的纤维和纺织材料。服装的穿着主要限制于人类，几乎是所有人类社会的共同特征。生理、社会和地理因素决定了服装穿着的数量和品种。一些服装样式有性别区分，但这种区分对跨着装者不适用。

1. 服装、身体和时尚

　　没有理由认为人的身体是禁止谈论的话题。之所以害怕，似乎是因为身体携带着某些关联性涵义。毕竟身体已经被严严实实地包裹了几百年，直到近代才被揭开。这个过程一旦开始，就快速地发展。身体越来越暴露引起了震惊，而且这种现象仍然遭受很大的怀疑。

　　然而，必须承认，我们将很多自然方面的概念与人的身体对应起来。例如，我们说一座山的"脚"，一条河峡的"颈"，一棵"站立"的树。米开朗基罗（Michelangelo）对此有更深刻的阐述，"人如果不精通人的身体，尤其是身体的解剖，就永远不能理解建筑的含义。"

　　我们所有比例概念都与身体相关。运动使我们的视野拓宽了，也使我们肌肉反应更加灵敏。不管睡着还是走着，我们对节奏的情感与我们规律的心跳和呼吸的起伏密切相关。

　　服装只有穿在身体上时才有意义。当它们被悬挂在衣橱里时显得极其无助。它们似乎在无声地怒吼，残忍的裁缝逼迫它们处于悲伤孤独的境地。要想真正了解服装，必须先了解它们背后的原因，然后再了解它们，只有这样才是可行的。服装不仅仅是纺织工厂的产品，或博物馆展示的物品，它们是人工制品，人们在日常生活的各种活动中都用到它们——站立、坐着、跳舞、工作或临终。只有当我们思考它们与身体的关系如何，怎样适应身体的时候，

专业词汇

1. clothing *n*. 服装
2. fiber *n*. 纤维
3. textile *n*. 纺织品
4. wear *vi*. 穿着
5. dresser *n*. 衣着者
6. fashion *n*. 时尚
7. proportion *n*. 比例
8. rhythm *n*. 节奏
9. wardrobe *n*. 衣橱
10. apparent *adj*. 易看见的
11. suit *vt*. 适合
12. style *n*. 风格

通用词汇

① assiduously *adv*. 勤勉地
② suspicion *n*. 怀疑
③ isthmus *n*. 地峡
④ pathetically *adv*. 哀婉动人的
⑤ artefact *n*. 人工制品
⑥ perverse *adj*. 堕落的

But the determining factor is really neither the body nor the clothes. It is Fashion. Fashion extends far beyond mere clothes; it is affected by how we stand, sit, smile and cry, how we love and how we hate. It is nothing less than an attempt to unify all the expressive capacities of language, gesture and physiognomy[7] in a given society.

Clothes represent an art form rising out of a period and environment and as such are no less valid than other artistic creations. Dress design[13] is a craft comparable with architecture, but lacking the latter permanence. It is an art which, like music, is in constant movement but, unlike music, cannot express direct emotion. As are both there arts, dress design is also non-figurative. It seeks not to pretend, but rather to display. It is at one and the same time fettered[8] and free, as is all genuine art. There is the simple type of dress which innocently declares its purpose, like a ball-dress[14], and then there are carefully constructed complex compositions like the armour[15] of the Middle Ages.

As do all other works of art, clothes reflect the times in which they were created. People reveal themselves unconsciously both in the art forms they accept, and those they reject. No form of art is more subtle than the variations that dress design creates upon that most fascinating of all themes—the human body.

When we consider the subject of dress through the centuries, it is not details of the dressmaker's[16] skill which concern us mod, but rather a gradual awareness of fashion as a camouflage, as a long series of device for hiding the true nature of men and women, created by God in his own image. The value of such duplicity[9] is surely negative, on a par with the conjuror who draws attention to his empty hand which does not do the trick.

The history of dress is full of deception[10] and self-delusion, deliberate mistakes and unconscious mistakes, the calculated and uncalculated in a long and illogical record of human folly. In short, it is simply the history of mankind. The interplay between the two sexes, with the whole range of shifting emotion that it encompasses, forms a mirror in which is reflected the ever-changing world of fashion. Dress does not merely show how men and women with to appear; it provides answers to many questions and also a criticism of the people who ask them. Women's fashion through the ages can provide a mass of silent evidence against the tyranny of the male at various times, and men's fashion can make out an equally strong case against the female sex. These situations occur because each sex reacts in accordance with the demands of the opposite sex. In a subtle way, the nature of one sex is revealed in the concealments of the other.

The history of dress cannot be treated as a conducted tour in which wheels revolve and carry both author and reader to a predetermined[11] destination. The reader must be a pedestrian[12] and use his legs; he must halt continually to adjust his bearings and change his point of view, sometimes to regard woman from the man's angle and at other times vice versa. It is necessary to observe and study all aspects of the history of dress from both these standpoints before we can hope to gain a real understanding of fashion.

它们的真正意义才变得明显。人的体型各种各样：瘦的、胖的；头大的、头小的。服装样式不能总是与身体框架一致，往往违背自然人体。如果身体不适合某种服装样式，那么应该修改服装，而不是身体。

但是真正的决定因素既不是身体也不是服装，而是时尚。时尚远远超出了单纯的服装，它影响我们如何站、坐、笑和哭，以及如何爱和恨。这无异于在一个特定的社会中，把语言、手势和相貌所有表达能力统一起来。

服装表现为某个时期和环境中产生的艺术形式，所以它不比其他艺术创造差。服装设计工艺可与建筑相比，但缺乏建筑的永久性。它是一种像音乐的艺术，不断运动；但又不像音乐，它不能直接表达情感。因为它们两者都是艺术，服装设计也是非具象，它不寻求伪装而是展示。与所有真正艺术一样，它将束缚和自由集于一体。有些服装构造简单，如舞会裙，很坦白地宣布它的目的，也有一些服装构造精心复杂，如中世纪的盔甲。

就像所有其他艺术作品那样，服装反映了它被创作的年代。人们无意识地揭示了他们接受和反对的艺术形式。没有哪种艺术形式像服装设计那样，用各种微妙的变化，为人体创建最迷人的主题。

当我们思考几个世纪以来的具体服装时，不是裁缝的细致手艺使我们想起时尚，而是渐渐对时尚有一种意识，时尚可以用作伪装，一种漫长的系列手法掩盖男性和女性的真实自然特征，即上帝根据自身形象创造的人。这种表里不一无疑起着负面作用，就像魔术师将人的注意力吸引到他没有做诡计的空手上。

着装历史充满了欺骗和自我欺骗，故意和无意犯错，精心策划或一时冲动，记录了长期以来人类愚蠢的和不符合逻辑的行为，简单地说，就是一部人类简史。男女之间的相互作用，包含了广泛范围内的情感变化，构成一面镜子，透过它映射出永远变化的时尚世界。服装不仅显示男女的外貌如何，还给予了很多问题答案，也同时批判提问题的人。在不同的历史时期，女性时尚提供了大量的反对男性独裁的无声证据。而男性时尚也对等地显示了强烈反对女性性别的情形。这些情况之所以发生是因为一种性别相应地对另一种性别的要求。当一种性别特征被揭示时，另一种性别特征就会以微妙的方式被掩盖。

对待服装历史不能像对待旅行团那样，转动车轮将作者和读者带到一个事先决定的目的地。读者必须是一个使用双腿的徒步者，不停地止步，调整他的态度和改变观点，有时从男性的角度看待女性，有时则相反。必须从男女两性观察和研究着装的历史，我们才有希望获得对时尚的真正了解。

专业词汇

13. design *vt.* 设计
14. ball-dress *n.* 舞会服
15. armour *n.* 盔甲，防弹衣
16. dressmaker *n.*（女装）裁缝

通用词汇

⑦ physiognomy *n.* 人相学
⑧ fetter *vt.* 束缚
⑨ duplicity *n.* 欺骗
⑩ deception *n.* 瞒骗
⑪ predetermine *vt.* 注定
⑫ pedestrian *n.* 行人

2. The uses of clothing

1) Utility

Clothing has evolved to meet many practical and protective purposes. The environment is hazardous, and the body needs to be kept at a mean temperature to ensure blood circulation and comfort. The bushman needs to keep cool, the fisherman to stay dry; the firemen needs protection from flames and the miner from harmful gases.

Humans have shown extreme inventiveness in devising clothing solutions to environmental hazards. Some examples include: space suits, air conditioned clothing, armour, diving suits, swimsuits[17], bee-keeper gear, motorcycle leathers[18], high-visibility clothing, and other pieces of protective clothing. Meanwhile, the distinction between clothing and protective equipment is not always clear-cut, since clothes designed to be fashionable often have protective value and clothes designed for function often consider fashion in their design.

2) Modesty

We need clothing to cover our nakedness[19]. Society demands propriety and has often passed sumptuary (clothing) laws to curb⑬ extravagance and uphold decorum⑭. Most people feel some insecurity about revealing their physical imperfections, especially as they grow older; clothing disguises and conceals our defects, whether real or imagined. Modesty⑮ is socially defined and varies among individuals, groups and societies, as well as over time.

In many Middle Eastern countries a debates still rages between liberals and fundamentalists as to how covered a women should be, and in many contemporary societies women still wear long skirts[20] as a matter of course. Europeans are generally less inhibited than Americans, but the trend for "casual Fridays" and dressing down office has been imported from the United States.

3) Sexual attraction[21]

Clothing can be used to accentuate⑯ the sexual attractiveness and availability of the wearer. The traditional role of women as passive sexual objects has contributed to the greater eroticization[22] of female clothing. Eveningwear[23] and lingerie[24] are made from fabrics that set off or simulate the texture[25] of skin. Accessories[26] and cosmetics[27] also enhance allure. Many fashion commentators and theorists have used a psychoanalytic⑰ approach, based on the writing of Sigmund Freud and Carl Jung, to explain the unconscious processes underlying changes in fashion.

The concept of the "shifting erogenous⑱ zone", developed by John Carl Flügel (1874—1955), a disciple⑲ of Freud in about 1930, proposes that fashion continuously stimulates sexual interest by cycling and focusing the attention on different parts of the body for seductive⑳ purposes, and that a great many articles of clothing are sexually symbolic of the male or female genitals. From time to time overtly sexualized clothing, such as codpiece[28] or the brassiere[29] comes into vogue[30].

2. 服装的使用

1）效用性

服装已经发展到能够满足很多实际和保护的用途。环境是危险的，身体需要保持均衡的温度以保证血液循环和舒适。布西曼人需要保持凉爽，渔夫需要保持干燥，消防员需要防火，矿工需要预防有毒气体。

人类在设计服装解决环境的危险性问题方面，已经显示出极大的创造性。一些例子包括：宇航服、空调服、防弹衣、潜水衣、游泳衣、防蜂服、摩托车皮服、高度可视服装和其他保护性服装。然而，服装和保护性设备的区别不总是很清晰，因为设计的时髦服装往往具有保护性能，设计的功能服装经常考虑时尚性。

2）遮羞性

我们需要服装覆盖我们的裸体。社会需要得体，经常颁布奢侈禁令限制奢侈和维持体统。大多数人感到显示自身身体的不完美有点不安全，特别是他们逐渐变老以后。服装伪装和掩盖我们真实或假想的缺陷。端庄是社会性定义，因不同的个体、群体和社会以及不同时代而变化。

在中东很多国家，在自由主义者和原教旨主义者之间仍然激烈地争论女性应该如何着装。在当代社会中，许多女性穿长裙被认为理所当然。欧洲人的着装限制渐渐地比美国人少，但"休闲星期五"和脱下办公服的潮流来源于美国。

3）性吸引

服装可用来强调穿着者的性吸引力和性能力。女性作为被动性对象的传统角色使女性服装更具性感。晚装和内衣采用的面料引起皮肤触感或模拟皮肤的肌理。配饰和化妆品也增强诱惑力。许多时尚评论家和理论家根据西格蒙德·弗洛伊德（Sigmund Freud）和卡尔·荣格（Carl Jung）的理论，采用了心理分析方法，解释时尚变化的潜意识过程。

"性感带转移"的概念由弗洛伊德的追随者约翰·卡尔·弗留葛尔（John Carl Flügel，1874—1955）大约于1930年提出，是指时尚不断地循环和聚焦身体的某个部位，激起性感兴趣，以达到诱惑目的，很多服装部件是男性或女性生殖器的象征。不时地有明显性感化服装成为流行时尚，如男性的遮阴布或女性的文胸。

专业词汇

17. swimsuit *n.* 泳衣
18. leather *n.* 皮革
19. nakedness *n.* 裸体
20. skirt *n.* 裙子
21. attraction *n.* 吸引

22. eroticization *n.* 性感
23. eveningwear *n.* 晚装
24. lingerie *n.* 贴身内衣
25. texture *n.* 纹理
26. accessory *n.* 配饰

27. cosmetic *n.* 化妆品
28. codpiece *n.* 遮阴布
29. brassiere *n.* 文胸
30. vogue *n.* 时尚
31. adornment *n.* 装饰

通用词汇

⑬ curb *vt.* 限制
⑭ decorum *n.* 端庄得体
⑮ modesty *n.* 端庄

⑯ accentuate *vt.* 强调
⑰ psychoanalytic *adj.* 心理分析的
⑱ erogenous *adj.* 唤起情欲的

⑲ disciple *n.* 信徒
⑳ seductive *adj.* 诱惑的

4) Adornment[31]

Adornment allows us enrich our physical attractions, assert our creativity and individuality, of signal membership or rank within a group or culture. Adornment can go against the needs for comfort, movement and health, as in foot-binding, the wearing of corsets[32] or piercing[33] and tattooing[34]. Adornments can be permanent or temporary, additions to or reductions of the human body. Cosmetics and body paint, jewelry[35], hairstyling and shaving[36], false nails[37], wigs[38] and hair extensions, suntans, high heels[39] and plastic surgery are all body adornments. People generally, and young women in particular, attempt to conform to the prevailing ideal of beauty. Bodily contortions[21] and reshaping through foundation garments[40], padding[41] and binding have altered the fashionable silhouette[42] throughout the ages.

5) Symbolic differentiation

People clothing to differentiate and recognize profession, religious affiliation[22], social standing or lifestyle. Occupational dress is an expression of authority and helps the wearer stand out in a crowd. The modest attire[43] of a nun announces her beliefs. In some countries, lawyers and barristers cover their everyday clothes with the garb[44] of silk[45] and periwig in order to convey the solemnity[23] of the law. The wearing of designer labels or insignias[24], and expensive materials and jewelry, may start as items of social distinction, but often trickle down through the strata[25] until they lose their potency as symbols of differentiation.

6) Social affiliation

People dress alike in order to belong to a group. Those who do not conform to the accepted styles are assumed to have divergent ideas and are ultimately mistrusted and excluded. Conversely, the fashion victim, who conforms without sensitivity to the rules of current style, in perceived as being desperate to belong and lacking in personality and taste. In some cases clothing is a statement of rebellion against society of fashion itself. Although punks[46] do not have a uniform[47], they can be recognized by a range of identifiers: torn clothes, bondage[48] items, safety pins[49], dramatic hairstyles[50] and so on.

7) Psychological self-enhancement

Although there is social pressure to be affiliated to a group, and many identical garments and fashions are manufactured and sold through vast chain stores, we rarely encounter two people dressed identically from head to toe[51]. While many young people shop with friends for help and advice, they do not buy the same outfits[52]. Whatever the situation, individuals will strive to assert their own personal identity through the use of make-up[53], hairstyling and accessories.

通用词汇

㉑ contortion *n.* 扭曲 ㉓ solemnity *n.* 庄严 ㉕ strata *n.* 社会阶层
㉒ affiliation *n.* 附属 ㉔ insignia *n.* 证章

4）装饰

装饰可以丰富我们身体的吸引力，体现我们的创造力和个性，象征某个团体或某种文化中的成员和地位。装饰可以违背舒适、运动和健康的需求，如裹脚、紧身胸衣、穿刺和文身。装饰可以是永久性或临时性的对身体进行添加或减少。化妆、体绘、首饰、发型、刮胡须、假指甲、假发、植发、晒黑、高跟鞋和整形手术等都是身体装饰。一般来说，人们尤其是年轻女性设法顺应流行的理想美。在历史上的不同年代，通过基础服装、衬垫和捆绑使身体扭曲和塑形，改变时尚廓形。

5）象征性差异

人们的着装可以区分和识别职业、宗教信仰、社会地位或生活方式。职业装是职权的表达，有利于穿着者在人群中突显。修女得体的服装宣告她的信仰。在一些国家，为了表达法律的尊严，律师和法律顾问每天穿丝绸长袍和戴假发。穿设计师品牌或名牌、昂贵材料和首饰，最初可以作为社会区分的物品，但是经常是由上向下流向社会低层，直到它们失去差别象征的功效。

6）社会附属关系

人们为了归属群体穿着相同服装。那些穿着与被认可样式不一致的人，被假定为具有不同的想法，最终不被信任，并被群体排除。相反，时尚牺牲者只会顺从，没有对现时时尚规律的敏感性，被认为极度渴望归属，缺乏个性和品味。在某些情形下，时尚服装本身是反抗社会的声明。尽管朋克没有制服，他们有很多被识别的特征：撕坏的服装、绑带、安全别针、奇特的发型等。

7）心理自我提升

尽管人们有隶属于某个群体的社会压力，使得许多相同的服装和时尚被制造出来，并在大量连锁店销售。但我们很少碰到两个人从头到脚穿得完全一样。许多年轻人和朋友一起逛商店，可以获得穿衣方面的帮助和建议，但他们不会购买相同的整套服装。不管什么情形，个人总是努力地通过化妆、发型和配饰来维护他们的个性特征。

专业词汇

32. corset *n.* 紧身胸衣
33. pierce *vt.* 刺穿
34. tattoo *n.* 文身
35. jewelry *n.* 首饰
36. shave *vt.* 剃须
37. nail *n.* 指甲
38. wig *n.* 假发
39. heel *n.* 后跟
40. garment *n.* 服装
41. pad *vt.* 衬垫
42. silhouette *n.* 轮廓
43. attire *n.* 服装
44. garb *n.* 装扮
45. silk *n.* 丝绸
46. punk *n.* 朋克
47. uniform *n.* 制服
48. bondage *n.* 绑带
49. pin *n.* 别针
50. hairstyle *n.* 发型
51. toe *n.* 脚趾
52. outfit 全套服装
53. make-up 化妆
54. apparel *n.* 服装

3. Different words

"Attire" means: 1. clothing of a distinctive style or for a particular occasion; 2: put on special clothes to appear particularly appealing and attractive.

"Apparel[54]" is very often a ceremonial type of clothing.

"Garment" refers to items of clothing in an almost official way-often linked to the word list when describing clothing belonging to someone.

"Garb" is akin to apparel and is very often associated with a form of clothing that is unusual or worn to disguise the person inside them.

"Clothing" is the simplest form of all to describe a collection of clothes. "Garment" and "apparel" are not identical. They may sometimes be used synonymously when used as adjectives, but when used as nouns they differ grammatically, for example:

—The word "garment" refers to a single piece of clothing. It is a countable noun.

—The word "apparel" refers collectively to clothing (and thus usually refers to more than one piece). It is an uncountable noun.

There are also differences in which words typically collocate with them when they modify[55] another noun. For example:

—a garment bag

—an apparel store

The term "costume[56]" can refer to wardrobe and dress in general, or to the distinctive style of dress of a particular people, class, or period. "Costume" may also refer to the artistic arrangement of accessories in a picture, statue, poem, or play, appropriate to the time, place, or other circumstances represented or described, or to a particular style of clothing worn to portray the wearer as a character or type of character other than their regular persona at a social event such as a masquerade[57], a fancy dress party or in an artistic theatrical performance.

One of the more prominent places people see costumes is in theatre, film and on television. In combination with other aspects, theatrical costumes can help actors portray characters' age, gender role, profession, social class, personality, ethnicity, and even information about the historical period/era, geographic location and time of day, as well as the season or weather of the theatrical performance. Often, stylized theatrical costumes can exaggerate some aspects of a character.

National costume or regional costume expresses local identity and emphasizes a culture's unique attributes. It is often a source of national pride. Examples of such are a Scotsman in a kilt[58] or a Japanese person in a kimono[59].

3. 不同的词

"Attire"意思：1. 独具风格或者特定场合的服装；2. 穿上特别的服装显得特有吸引力和魅力。

"Apparel"经常是指一种礼仪性服装。

"Garment"是一种稍微正式的用法，经常在列举服装属于某人时用这个词。

"Garb"与"Apparel"相似，通常指不寻常的服装形式，或穿上它们用来伪装穿着者。

"Clothing"是描述服装总称的词汇中形式最简单的词。"Garment"和"Apparel"不同之处在于，当用作形容词时它们是同义词，但是作为名词时，它们在语法上不同，例如：

"Garment"指单独一件衣服，是可数名词。

"Apparel"指服装的整体（因此，通常指服装多于一件），是不可数名词。

当它们与其他词搭配或修饰其他名词时，也不相同。

例如：一个"服装"袋，一家"服装"商店。

"Costume"一般指衣橱和着装，或者指特定的人、阶层或时期的特定风格的服装。"Costume"也可以指在绘画、雕像、诗歌或戏剧中，配件的艺术安排，或者与所表现或描绘的时间、地点或场景相吻合的服装；或者某种特定样式的服装，塑造穿着者在特定社会活动中具有某种特征或某类特征，而不是他们正常的个性，例如化妆舞会，一种幻想的着装聚会或一种艺术性戏剧表演。

"Costume"更多地用于剧院、电影和电视上的服装。与其他方面相结合，舞台服装可以帮助演员塑造人物年龄、性别角色、职业、社会阶层、个性、种族，甚至有助于体现戏剧表演中讲述的历史时期／年代、地理位置和时间、季节和天气。通常情况下，独具风格的舞台服装可以夸张某种特征的多个方面。

民族服装和地域服装表达了地域特征，强调了一种独特的文化性，往往是民族自豪感的源泉，例如苏格兰人穿褶裥短裙和日本人穿和服。

专业词汇

55. modify *vt.* 修改
56. costume *n.* 戏装
57. masquerade *n.* 化装舞会
58. kilt *n.* 褶裥短裙
59. kimono *n.* 和服

Passage 2

Fashion

The Oxford English Dictionary defines fashion as "current popular custom or style, especially in dress."

While everybody talks about fashion nowadays, hardly anyone ever stops to consider what truly is, what its origins are, or how it came to occupy all spheres① of society. In our consumerist culture, nothing escapes its influence, and it is therefore safe to say that today fashion has become a way of life.

Conceived② in the context of dress, fashion as a logic bases on novelty③ has extended to all areas of society, a fact first confirmed in 1890, when the French Sociologist Gabriel de Tarde defined it as a social process independent of dress. Homo sapiens④ are the only animals that wear clothes, and fashion came into being because men and women are social animals who, while desiring to belong to a group, also want to be different, as pointed out by German sociologist George Simmel. Defining fashion is not easy, because fashion is multifaceted⑤. It forms part of the culture and thus can be studied from multiple⑥ angles[1] from the perspective of history, sociology, anthropology, psychology, art, economics, or science. Fashion is a complex process that reflects society's transformations in each age.

Fashion has been influenced by wars, conquests, laws, religion, and the arts. Individual personalities have also had an impact on fashion. Fashion is now directly linked with film, music, literature, arts, sports and lifestyle as never before. The contribution of fashion and its growing influence has also permeated⑦ into other aspects of the business sector as has never before been witnessed. Ever fashion follows a cycle, and fashion cycle has no specific measurable time period. Some styles sustain⑧ for longer period or some die out soon and some styles come back years after it was declined⑨. So we can say fashion changes with time and has always been evolving to fit the taste, lifestyle and demands of society.

专业词汇

1. angle *n.* 角度

第 2 课

时　尚

　　牛津英语词典定义时尚为"正在流行的风俗或样式，尤其指着装"。
　　今天虽然每个人都在谈论时尚，但是几乎没有人停下来思考什么是真正的时尚，时尚源于何处，为什么时尚占据社会各个领域？在我们用户至上的文化中，什么事都逃不过时尚的影响，因此可以很有把握地说，今天时尚已经成为一种生活方式。
　　在服装领域内构想的以新奇为逻辑基础的时尚，已经延伸到社会各个领域，首先被确认的一个事实是，在1890年法国社会学家加布里埃尔·塔尔德定义时尚为与服装独立的一个社会过程。德国社会学家格奥尔格·齐美尔指出，所有动物中只有智人（现代人的学名，译者注）穿衣服。时尚的出现是因为男性和女性是社会性动物，他们既希望隶属于一个群体，又希望与其他人不同。定义时尚很难，因为时尚具有多面性，它构成文化的一部分，因此可以从历史学、社会学、人类学、心理学、艺术、经济学和科学等多个角度地去研究它。时尚是一个复杂的过程，它反映了各个时期社会的变革。
　　时尚曾受到战争、征服、法律、宗教和艺术的影响。个人性格也对时尚产生影响。以前的时尚从未像现在这样直接与电影、音乐、文学、艺术、体育和生活方式联系在一起。时尚的贡献和它逐渐增大的影响力已经渗透到各行各业，这也是以前从未目睹的。时尚永远遵循周期，时尚周期没有具体可测量的时间段。有些样式持续较长时期，有些很快消亡；而另一些时尚在它衰落以后又重新回归。因此，我们可以说时尚随时间而变化，并一直不断地发展，以适应品味、生活方式和社会需求。

通用词汇

① sphere *n.* 球
② conceive *vt.* 构思
③ novelty *n.* 新奇
④ Homo sapiens *n.* 智人
⑤ multifaceted *adj.* 多方面的
⑥ multiple *adj.* 多重的
⑦ permeate *vt.* 弥漫
⑧ sustain *vt.* 维持
⑨ decline *vi.* 下降

1. The origin and evolution of fashion

Fashion is a particular system of production[2] and organization of dress that emerged in the West with the advent of modernity during the fourteenth century, subsequently[10] expanding with the rise of mercantile[11] capitalism, hand-in-hand with technological processes. Pre-modern societies were traditional-based[3] on worship of the past, of tradition-perpetuating the same forms of dress with negligible[12] alterations[13].

The system of fashion took root when a rupture[14] from the past (from the old) in benefit of the future (the new) occurred, which is to say, when newness became a constant and general principle[4], highlighting a predilection[15] typical of the West: modernity.

There is in fashion a vital trait of modernity: the abolition of traditions. Fashion has a characteristic of the modern because it is an indication of emancipation[16] from, among other things, authorities. Fashion is irrational[17]. It consists of change for the sake of change, whereas the self-image of modernity consisted in there being a change that led towards increasingly rational self-determination.

Modernization consists of a dual movement: emancipation always involves the introduction of a form of coercion[18], since the opening of one form of self-realization always closes another. Every new fashion is a refusal to inherit[19], subversion against the oppression of the preceding fashion. Seen in this way, emancipation lies in the new fashion, as one is liberated from the old one. The principle of fashion is to create an ever-increasing velocity[20], to make an object superfluous as fast as possible, in order to let a new one have a chance.

Fashion is irrational in the sense that it seeks change for the sake of change, not in order to improve the object, for example by making it more functional[5]. It seeks superficial changes that in reality have no other assignment than to make the object superfluous on the basis of non-essential qualities, such as the number of buttons[6] on a suit jacket[7] or the famous skirt length[8]. Why do skirts become shorter? Because they have been long. Why do they become long? Because they have been short. The same applies to all other objects of fashion.

The evolution of the fashion system can be divided into three stages:

—Aristocratic[21] fashion appeared in the second half of the fourteenth century and lasted until the middle of the nineteenth century. Its dominant figure[9] was masculine[10], with men exhibiting the full range of their power through a fashion based on ornamentation[11].

—Centennial fashion emerged in the second half of nineteenth century and extended up to the 1960s. Men were eclipsed by women, who drew attention to themselves with haute couture[12] designs.

—Open fashion was born in the 1960s and continues to this day, characterized by the great interest of both sexes in their appearance, coinciding with the rise of consumer society.

1. 时尚的起源和演变

时尚是服装生产和组织的一种特别的体系，在14世纪西方工业化来临之际出现，随后在兴起的商业资本主义和伴随的技术进步过程中逐步发展。社会工业化之前，人们尊崇过去的传统，也尊崇变化可以忽略不计的、相同形式的传统服装。

时尚体系要与过去（旧的）彻底决裂，有利于未来（新的）产生。即当"新"成为一种持续的普遍原理时，就突显出西方一种典型的偏爱：现代性。

时尚现代性的重要特征：废除传统。时尚具有现代性特征，是因为它表明了要从权威和其他事项中解放出来。时尚是非理性的，它为了变化而由变化构成，因为现代性的自我形象就是以变化而存在，是日趋理性的自我决定性的变化。

现代化由一种双重运动构成：解放总是涉及采用高压形式，因为一种自我实现形式的开始，总是要结束另一种形式。每一种新时尚就是拒绝继承和颠覆先前时尚的压迫。从这个角度看，解放在于新时尚，因为新的从旧的中被解放出来。时尚的原理就是创造不断递增速度，使一种事物尽快地过剩，让新的获得机会。

时尚的非理性含义是为变化的缘故而寻求变化，而不是为了改进事物，例如不是使事物更具功能性。它试图改头换面，而事实上是改变非本质性特征，使事物成为多余，例如改变夹克套装上纽扣的数量或著名的裙摆长度。为什么裙子变短？因为它们一直是长的。为什么变长？因为它们一直是短的。这同样适用于其他所有时尚事物。

时尚体系的演变分为三个阶段：

——贵族时尚，产生于14世纪下半叶，并持续到19世纪中叶。主导形象是男性化，以装饰为基础的男性时尚展示他们全方位的权力。

——百年时尚，产生于19世纪下半叶，并持续到1960年代。女性自身关注高级时装设计，使男性黯然失色。

——开放性时尚，产生于1960年代，并一直持续到今天。其特点是男女都对其外观有极大兴趣，与消费社会的兴起相吻合。

专业词汇

2. production *n.* 产品
3. tradition *n.* 传统
4. principle *n.* 基本原理
5. function *n.* 功能
6. button *n.* 纽扣
7. jacket *n.* 夹克
8. length *n.* 衣长
9. figure *n.* 体型
10. masculine *adj.* 男性气质
11. ornamentation *n.* 装饰
12. haute couture *n.* 高级时装

通用词汇

⑩ subsequently *adv.* 其后
⑪ mercantile *adj.* 贸易的
⑫ negligible *adj.* 可以忽略的
⑬ alteration *n.* 变化
⑭ rupture *n.* 断裂
⑮ predilection *n.* 偏爱
⑯ emancipation *n.* 解放
⑰ irrational *adj.* 不合理的
⑱ coercion *n.* 强迫
⑲ inherit *vi.* 继承
⑳ velocity *n.* 速率
㉑ aristocratic *adj.* 贵族的

2. Fads[13] and fashions

In academic and popular discussions, fads and fashions are often treated together. They help fill in large culturally blank areas that haven't explained with other forms of collective obsessions[22]. Fads and fashions occur within nearly every sphere of social life in modern society, most obviously in the areas of clothing and personal adornment. The line separating fads and fashion is hardly clear, as both terms are frequently applied to aspects of change in the physical[14] presentation of the self and to areas not involving large economic investments.

Fads and fashion can occur together and should be understood as expressive rather than instrumental[23] forms of collective behavior. Because they are noninstrumental actions they show a high degree of emotional involvement. Fads and fashions differ as well: fads are more spontaneous and tend to not follow the cycles that fashions do.

Fads are temporary, highly imitated outbreaks of mildly unconventional[15] behavior. In contrast a fashion is a somewhat long-lasting style of imitative behavior or appearance. A fashion reflects a tension between people's desires to be different and their desire to conform. Its very success undermines its attractiveness, so the eventual fate of all fashions is to become unfashionable. Fads also more frequently involve crowds and face-to-face interaction, whereas fashion usually involves diffuse[16] collectivity, in which widely dispersed individuals respond in a similar way to a common object of attention.

Fashion is much like fads and other collective obsessions, except that it is institutionalized and regularized, becoming continuous rather than sporadic[24], and partially predictable[17]. Whereas fads often emerge from the lower echelons[25] of society, and thus constitute a potential challenge to the class structure of society, fashion generally flows from the higher levels to the lower levels, providing a continuous verification of class differences. Continuous change is essential if the higher classes are to maintain their distinctiveness[26] after copies of their clothing styles appear at lower levels. Thus, fads and fashions contribute to both social integration and social differentiation. With fashions tending to change cyclically within limits set by the stable culture. For fads and fashions, established groups usually serve as the settings and conduits through which the behavior passes.

Fashion trends[18] change every season, bringing into trend diverse styles which can make you look fabulous. Fashion plays a very important role when it comes to the way others perceive us so paying as much attention to style can have a positive impact over the physical appearance. Fashion offers us new trends to choose from every season, allowing diversity to take over, avoiding this way a fashion routine.

3. Fashion cycle

The way fashion change is described as fashion cycle. Cycle means period of time or life span during which fashion exists.

2. 热潮和时尚

在学术和大众讨论中，热潮和时尚经常放在一起探讨。它们很好地解释了集体着迷现象，此外没有其他形式能够解释这种现象，大大填补了文化上的空白区域。在现代社会，几乎各个社会生活领域都有热潮和时尚的发生，服装和个人装饰领域最为明显。热潮和时尚之间没有明确的分界线，因为两个词汇经常用于个人身体外表的变化方面，涉及的范围没有大的经济投入。

热潮和时尚可以一起产生，应该理解为集体行为的表达形式，而非制度化形式。因为它们是非制度化行为，显示出高度的情感成分。热潮和时尚也有区别：热潮较有自发性，不像时尚那样遵循周期。

热潮是短暂的、突发的、高度模仿的和温和的非常规行为。相比之下，时尚对行为和外貌的模仿持续的时间长一些。时尚反映了人们既希望与别人不同，又希望与别人趋同的矛盾。时尚非常成功地削弱自身的吸引力，因此所有时尚的最终命运是变得不时尚。热潮经常涉及大量人群和面对面的相互作用，而时尚通常涉及集体的传播，是广泛分散的个体关注一个共同的对象，以相似的方式做出反应。

时尚很像热潮和其他的群体痴迷现象，但它被制度化和系统化，使其成为持续的而不是断续的，部分地可预测。然而，热潮经常在社会的下层出现，因此，潜在地构成了对社会等级结构的挑战。时尚一般是从高层流向低层，提供了阶级差别的持续验证。如果高层阶级的服装样式被低层阶级模仿后，高层阶级要维持他们的独特性，那么不断变化就是必要的。因此，热潮和时尚有利于社会的整合和社会差别化。稳定的文化设定了时尚只能在有限的范围内周期性变化。在一些固有的群体中，热潮和时尚通常起着实现行为的环境和渠道作用。

时尚潮流每个季节在变化，带来了丰富多彩时尚样式，可以让你焕然一新。当提到别人看待我们的方式时，时尚起着十分重要的作用。因此高度注意风格，对我们生理外貌有积极的影响。时尚为我们提供了多种新的潮流，供我们从每个季节中选择，使差异化产生，避免了一种时尚路线的方式。

3. 时尚周期

时尚变化的方式可用时尚周期来描述。周期是指时尚存活的时间段或寿命。

专业词汇

13. fad *n.* 狂潮
14. physical *adj.* 身体的
15. unconventional *adj.* 非传统
16. diffuse *adj.* 传播
17. predictable *adj.* 可预测
18. trend *n.* 趋势

通用词汇

㉒ obsession *n.* 着魔
㉓ instrumental *adj.* 起作用的
㉔ sporadic *adj.* 不定时发生的
㉕ echelon *n.* 等级
㉖ distinctive *adj.* 有特色的

1) The stages of the fashion cycle

The fashion cycle is usually depicted as a bell-shaped[19] curve[20] encompassing five stages: introduction, increase in popularity, peak in popularity, decline in popularity, and rejection. Consumers are exposed very season to a multitudes of new styles created by designers or launched[21] by big clothing brands[22]. It is seem some styles are rejected immediately by the buyers[23] on retail[24] level, where as some styles are accepted for a time, as demonstrated by consumers purchasing and wearing them. With trend reports in new papers and fashion channels showing latest trends many women who consider themselves fashionable, or up to date with what's new, go out each season to assess what's needed in order to keep her wardrobe relevant. Then designers also are constantly going back in time for inspiration[25]. Each season a new version[26] of the old era is tapped and we see a few small changes to looks that have all walked down the catwalks[27] before.

(1) Introduction of a style

Every designer each season works on a new collection[28], interpret their research into apparel. Every style has some different elements[29] like line[30], shape[31], colour[32], fabric. The first stage of the cycle where the new style is introduced may or may not be accepted by the consumers. Every style is reviewed at design centre and in fashion shows[33]. New styles are usually introduced in high price level. Usually a new style created by a designer is worn by the selected people who can afford it, and mostly these people are fashion leaders like celebrities[34] and rich people who loves to experiment and try out new styles to grab the attention of media. Such styles as they are expensive are produced in a small quantity.

(2) Increase in popularity

A new style worn by a celebrity or famous personality, seen by many people and it may draw attention of buyers, the press, and the public. Most designers also have prêt-à-porter[35] line that sells at comparatively low prices and can sell their designs in quantities. Manufacturers adopt design and styles to produce with less expensive fabric or less details[36]. The adaptations are made for mass production.

(3) Peak in popularity

Styles at this stage are most popular. When production of any style is in volume[37], it requires mass acceptance. The manufacturers carefully study trends because the consumer will always prefer clothes that are in the main stream[38] of fashion. When a fashion is at height of its popularity, it may be in such demand that many manufacturers copy it or produce adaptations of it at many price levels. Length at this stage determines if the fashion becomes classic or fad.

(4) Decline in popularity

A time comes after the mass production of a few styles people get tired and began looking for new styles. They still wear the particular style but are not willing to buy them at the same price. With the launch of new collection every season the popularity of the style of the previous seasons declines. Fashion is over saturated or flooded the market. Retail stores put such decline styles on sale rake as off season sale or clearing sale.

1）时尚周期的不同阶段

时尚周期通常用铃形曲线来描绘，包含 5 个阶段：引入、上升、高峰、下降和拒绝。每一个季节消费者可以看到很多新的款式，它们由设计师设计或由大的服装品牌公司推出。一些款式似乎在零售层面上立刻被购买者拒绝，而另一些款式被接受的时间长一些，它由消费者的购买和穿着来证明。那些认为自己很时髦的女性，或认为紧跟新潮的女性，为了使她们的衣柜紧跟潮流，必须根据报纸报道最新趋势，时尚频道播放最新潮流，评估每个季节的新潮中自己需要什么。而设计师却要不断地回到过去寻找灵感。过去的款式往往被开发成新款，当在 T 台上展示时，我们看到了一些小的改动。

（1）样式的引入阶段

每个设计师都要为每个季节的新产品工作，将他们的研究演绎成服装。每种样式有一些不同的元素，例如线条、形状、色彩和面料。在周期的第一阶段，一种新样式被引入时消费者可能接受或可能不接受。每一种样式在设计中心和时尚秀中被审查。新样式在引入阶段通常价格很高。设计师设计的新样式通常针对那些买得起的人，这些人大多是时尚领袖，如名人和富人，他们喜欢体验和试穿新样式，吸引媒体的注意。这些新样式因昂贵被少量生产。

（2）人数增长阶段

当很多人看到明星或名人穿新样式的服装时，可能会吸引买手、新闻媒体和公众的注意。大多数设计师还有成衣生产线，以相对低的价格销售，可以批量销售他们的设计。生产商采纳设计和样式，用便宜的面料和较少的细节生产，这是适应大众的产品。

（3）人数高峰阶段

样式在这个阶段是最流行的。任何样式大批量生产时，需要大众接受。制造商仔细研究趋势，因为消费者总是喜欢处于时尚主流阶段的服装。当一种时尚处于流行的最高峰时，就会出现这种需求，很多制造商复制它或者稍作改动以不同的价格销售。这一时段的长度决定了此种时尚变成经典还是变成热潮。

（4）人数下降阶段

一些样式在成为大众产品的一段时间以后，人们厌倦了，并且开始寻求新的样式。他们仍然穿着那些样式，但是不愿意再以相同的价格购买它们。由于每一季节有新的样式推出，前几季节流行的样式就衰落了。饱和或洪水般的时尚市场就此结束。零售商店将这些过时的服装打折或清仓处理销售。

专业词汇

19. bell-shaped 钟形
20. curve *n.* 弧线
21. launch *vt.* 发布
22. brand *n.* 品牌
23. buyer *n.* 买手
24. retail *n.* 零售
25. inspiration *n.* 灵感
26. version *n.* 版本
27. catwalk *n.* T 形台
28. collection *n.* 系列时装
29. element *n.* 元素
30. line *n.* 线条
31. shape *n.* 形状
32. colour *n.* 色彩
33. show *n.* 发布会
34. celebrity *n.* 名人
35. prêt-à-porter 成衣
36. detail *n.* 细节
37. volume *n.* 体积
38. stream *n.* 潮流

(5) Rejection[27] period

It is the last phase of the cycle. Some consumers have already turned to new looks, thus beginning a new cycle. The rejection or discarding of a style just because it is out of fashion is called consumer obsolescence. Since consumers are no more interested manufactures stop producing the same and the retailers will not restock the same styles. Now it's time for a new cycle to begin.

Figure 2.1   Stages of fashion cycle
图 2.1   时尚周期的不同阶段

2) Length of the cycles or fashion movement

Fashion cycle has no specific measurable time period. Ongoing motion of fashion in the fashion cycle is a movement. Fashion movement is affected by

• Economic or social factors
• Invention of fibers or fabrics
• Advertising of the product

Rate of the movement varies with each fashion. Short time to peak in popularity, other takes longer ; Some declines slowly, others swiftly.

(1) Classics

Fashion that always remains in the rise stage of the fashion cycle is known as classic.

The styles that remain more or less accepted for an extended period. These styles never become completely obsolete. Example: Classic shirt, Jeans[39] and Tailored suit.

(2) Fad

Fad also knows as short-lived fashion and can hold the attention of the consumer for a very short period. The consumer group is very small and the garments are low priced and flood the market in very short time. The consumer gets tired of the designs due to market saturation and they die out soon.

Figure 2.2   Cycles for fad and classic, compared with normal fashion cycle (Solid line represents normal cycle)
图 2.2   热潮和经典的周期，与正常时尚周期对比
（实线表示正常时尚周期）

（5）拒绝阶段

这是周期的最后阶段。一些消费者已经转移到新样式上，从而开始一个新的周期。一种样式的拒绝或丢弃，只是因为它过时了，被称为消费过时。因为消费者不再感兴趣，制造商就停止生产这种样式，零售商也不再储存这种样式。现在是到了新周期开始的时候了。

2）周期长度或时尚运动

时尚周期没有具体的可测量的时间段。时尚周期中正在运行的动势是一种运动。时尚运动受以下因素的影响：

- 经济或社会因素
- 纤维或面料创新
- 产品广告

每种时尚的运动速率不同，一些时尚在很短的时间内到达顶峰状态，而另一些时尚需要很长时间到达高峰；一些时尚在很长时间内衰落，而另一些在很短时间就衰落。

（1）经典

总是保持在时尚周期上升阶段的时尚，被称之为经典。仍然有一些样式在长时间段里或多或少地被人们接受，这种样式永远不会变得完全过时，例如经典衬衫、牛仔和西装。

（2）热潮

热潮是指短生命周期的时尚，在相当短的时间内吸引消费者的注意力。消费群体非常小，服装价格很低，在很短时间内充斥市场。由于市场饱和，消费者厌倦这种设计，因此它们很快消亡。

专业词汇

39. jean *n.* 牛仔裤

通用词汇

㉗ rejection *n.* 拒绝

3) Recurring cycles

Fashion designers draw inspirations from past. It has been noted that styles reappears years later and is reinterpreted for a new time. A change in element is normal like change in the silhouette or proportion may recur and is sometimes interpreted with a change in fabric and detail. Many nostalgic looks are drawn by designers from the 1940s, 50s, 60s, 70s & 80s. However, the use of different fabrics, colours, and details make the looks unique in every creation.

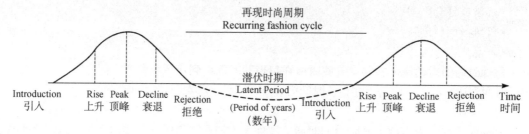

Figure 2.3 Recurring fashion cycle
图 2.3 再现时尚周期

3）周期复苏

时尚设计师从过去汲取灵感。已经注意到，样式在多年后又再次出现，在新时期得到重新诠释。一般在轮廓或比例的要素方面作改变，有时通过面料和细节诠释变化。许多怀旧的样式就是设计师回归20世纪40年代、50年代、60年代、70年代和80年代样式，然而由于使用了不同的面料、色彩和细节，每一种设计呈现出独特的样貌。

Passage 3

The Types of Fashion

The fashion industry is divided into five main markets according to price point: Haute couture, designer, bridge, moderate① and mass. However, there are additional markets that are just as important to be aware of, including one-of-a-kind, bespoke[1], contemporary, secondary, private label, and discount. The following provide a listing and explanation of all of the fashion industry markets, from highest to lowest price point.

1. One-of-a-kind

A one-of-a-kind piece or ensemble[2] is fully customized and made-to-order[3] for a specific client according to his or her exact measurements[4] and specifications②. One-of-a-kind garments are considered the pinnacle③ of luxury[5] in the fashion world because only one of its kinds is in existence. Custom-made[6] garments are crafted at the haute couture level, using only the finest fabrics, trims[7], embroideries[8], and appliques[9]. The price point reflects that level, due to the quality of materials used and the superior extent of detail and workmanship that goes into making each piece. Custom clothing is often referred to as the *pièce de resisitance* because it is considered a true, irresistible④ showpiece at every level. It is considered by many to be an art form[10]. Finished custom pieces are often displayed in museum exhibits around the world and sell for thousands of dollars at auction. A custom client may request one piece or an entire wardrobe for a series of special events, such as black-tie[11] galas. It is the responsibility of the designer to come up with each of those items according to a specified timeline and perhaps a personal branding theme[12] that the client wishes to be carried out throughout his or her customized ensemble.

Celebrities who are presenters at an awards show, or who have received industry award nominations, will often be seen wearing a custom dress designed especially for the occasion. Other custom clients may include a celebutante (a person who is famous for being famous), a jet setter or socialite who is attending an exclusive event. A debutante⑤ who is making her debut into society at the cotillion⑥ ball, a high-profile businesswoman who is being honored at a conference, a low-profile client who prefers to remain anonymous after receiving an inheritance, or anyone who has an appreciation for custom clothing.

第 3 课

时装类型

时装行业按照价位分为五个主要市场：高级时装、设计师品牌时装、过渡类服装、中等类服装和大众型服装。但是，还有其他一些重要市场不能忽视，包括一对一、定制、流行、次品牌、私有品牌和打折服装。下面提供了时装行业从最高到最低价位所有市场类型的列表和解释。

1. 独一无二类型

独一无二类型的服装或套装是一种唯一型时装，根据特定顾客，他或她的准确尺寸和规格，完全客户化定制。独一无二的服装被看成是时装界最奢侈的，因为它是同类中的唯一一件，客户化定制的服装以高级时装工艺水准制作，只用最好的面料、辅料、刺绣和珠绣。价位反映了它使用面料的质量、超细的细节和每一件所需要的手工。定制服装经常被看成是一件超级精品，因为在各个方面它都被看成是一件真正的、诱惑人的展品，很多人认为是一种艺术形式。定制服装的成品经常在世界各地的博物馆里巡回展出，在拍卖会上卖几千美元。定制客户可能需要一件或一整套定制服装，去参加一系列特别活动，例如黑领带半正式礼服。设计师的责任是根据明确规定的时间表完成每一件单品，也许客户定制服装里隐含客户希望的个性品牌理念。

在颁奖典礼上做主持的名人，或获得行业提名的人，经常看到他们在这种场合穿定制的服装。其他定制客户还包括那些频频出现在镜头前的名媛（出名是因为出生名门）、阔佬或上流社会人士，他们要参加一个特别活动。初进社交界的上流社会年轻女子初次参加沙龙舞会，一个成功职业女性很荣幸参加某个会议，一个身份低的女性在得到了遗产后仍想保持匿名，

专业词汇

1. bespoke *adj.* 定制
2. ensemble *n.* 全套服装
3. made-to-order 定做
4. measurement *n.* 度量
5. luxury *n.* 奢侈
6. custom-made 定制
7. trim *n.* 装饰
8. embroidery *n.* 刺绣
9. applique *n.* 缝饰
10. form *n.* 形式
11. black-tie 男子半正式礼服
12. theme *n.* 主题

通用词汇

① moderate *adj.* 中等的
② specification *n.* 规格
③ pinnacle *n.* 顶点
④ irresistible *adj.* 无法抗拒的
⑤ debutante *n.* 初进社交界上流社会年轻女子
⑥ cotillion *n.* 沙龙舞

Costume designers are fashion designers who design and create customized costumes for film, televise, performing arts and stage productions, fashion shows, special events, or other performances for "talent" or show business personalities, actors, models, singers, dancers, and other performers. The process sometimes involves extensive research of a historical component⑦, such us the replication⑧ of clothing from a particular era, needs to be reproduced. Once the research is complete, designs are sketched, and fabric is sourced and purchased, then draped on a form (i.e., mannequin[13]) or patterned and then produced. The costumes oftentimes require accessories, such as hats[14], headdresses[15], tiaras[16] and other jewelry, hosiery[17], masks[18], wigs, and footwear. The process may involve the creation of something unique, like a full-body cat suit for a musical. Costume designer John Napier won a Tony Award in 1983 for Best Costume Designer, for the Broadway musical Cats. A singer such as Britney Spears will need a completely customized wardrobe created for her worldwide concert tours, consisting of several head-to-toe outfits for each series of songs, matching each corresponding stage set. So the costume designer will need to carry out a feeling in the costumes and ensembles that will correspond⑨ with the overall concert theme. Singer and actress Madonna had 85 costume changes in the movie Evita, which shows how important a role a costume designer plays in the overall production of a movie.

Certain unique circumstances come into place for a costume designer that an ordinary fashion designer would not necessarily encounter. For example, costume designers have to pay special attention to the needs of the particular person they are fitting. For a dancer, the fit of his or her clothing is critical in ensuring that movement is not inhibited during performances.

2. Haute Couture

French for "high sewing", or "high fashion", haute couture (often referred to more informally as "couture"), describes handmade, made-to-measure[19] garments using only the most luxurious fabrics, such as the finest cashmere[20], fur[21], suede[22], leather, and silk, sewn with extreme attention to detail by the most skilled seamstresses[23], often using hand-executed techniques. It is the fusion[24] of both costume and high fashion and is often seen on the most affluent⑩ and famous people.

The *Chambre Syndicale de la Haute Couture* is an association whose members include those companies that have been designated to operate as an haute couture atelie⑪ or house. Haute couture is a legally protected and controlled label and can only be used by those fashion houses they have been granted this designation by the French ministry of industry. This governing body annually reviews its membership base, which must comply with a strict level of regulations and standards in order to maintain membership. The membership list changes annually as a result of its stringent⑫ criteria.

通用词汇

⑦ component *n.* 成分
⑧ replication *n.* 复制
⑨ correspond *vi.* 相应
⑩ affluent *adj.* 富裕的
⑪ atelier *n.* 工作室
⑫ stringent *adj.* 严格的

或是任何想定制服装的那些客户。

戏装设计师是时装设计师，为电影、电视、表演艺术和舞台产品、时装发布会、特殊活动，或为其他表演"天才"或显示职业个性的演员、模特、歌手、舞蹈者，以及其他表演者设计和制作定制化服装。有时这个过程要对涉及的历史因素进行广泛的研究，然后再现它们，例如对某个特定时期服装的复制。研究一旦完成，设计草图，寻找和购买面料，在人台上（人体模型）进行设计或者画纸样，然后生产。这种服装经常需要配饰，例如帽子、头巾、冠状头饰和其它首饰，袜子、面具和鞋类。这个过程也许包括一些独特的设计，例如为《猫》音乐剧设计套装。戏装设计师约翰·纳皮尔（John Napier）在1983年为百老汇音乐剧《猫》设计的服装获得了最佳戏剧服装设计师托尼（Tony）奖。例如歌手布兰妮·斯皮尔斯（Britney Spears）为世界巡回音乐会就需要设计一整套定制的服装。从头到脚需要好几套服装，与每一首歌曲设置的舞台背景相吻合。因此，戏装设计师需要在服装中夹带着情感，与整个音乐会主题相吻合。歌手和演员麦当娜（Madonna）在电影《贝隆夫人（Evita）》中更换了85次服装，这就显示出戏装设计师在整个电影作品中起着极其重要的作用。

对戏装设计师来说，需要再现某种特定场景，而对时装设计师来说，就不会碰到这个问题。例如，戏装设计师必须特别注意设计服装适合特定的人物。一个舞蹈者，他或她的服装在表演时要确保能够活动自如，而不能受到压制。

2. 高级时装

"Haute Couture"法语高级缝制或高级时装（不正式使用时直接称"Couture"），比喻手工制作、量体裁制的服装，仅使用最昂贵的面料，例如上等羊毛面料、裘皮、小山羊皮、皮革和丝绸，由最熟练的缝纫师精工细作，经常使用手工技术。它是戏装和高级时装的融合，经常看到最富有和最有名的人穿着。

巴黎时装协会由已经获得高级时装工作室或时装屋的资格成员构成。高级时装是受法律保护和受控制的品牌，只有那些被法国行业部授权了的时装屋才能使用。这个管理机构每年审查它的成员基地，基地必须严格遵守规章和标准，以便维持会员资格。由于标准严格，成员名单每年都在变化。

专业词汇

13. mannequin *n.* 人体模型
14. hat *n.* 帽子
15. headdress *n.* 头饰
16. tiara *n.* 冠状头饰
17. hosiery *n.* 袜类
18. mask *n.* 面具
19. made-to-measure 量体裁制
20. cashmere *n.* 山羊绒
21. fur *n.* 毛皮
22. suede *n.* 小山羊皮
23. seamstress *n.* 女裁缝
24. fusion *n.* 融化

The couture house is headed by a fashion couturier who oversees a workroom of skilled workers who practice their hand-made crafts[25] as experts in whether dressmaking or tailoring. The process may begin with a sketch, an illustration, or a draped and cut muslin[26] or toile[27], depending on the designer's preference. To finalize a couture piece, fine trim, embroidery, and embellishments[28] are often purchased by outside sources, who are expert practitioners[13] in their respective field and then meticulously sewn into each piece. Exquisite fit is an inherent[14] quality of couture piece. The client will endure a series of fittings to determine that exact measurements have been achieved, to ensure not only precise fit but also style and comfort, which are equally essential.

When haute couture collections were first produced, they were presented to the press, buyers, and high-end clientele in a trunk show[29] format in a designated salon. Each model carried a card that indicated a corresponding look number, making it easy for those in attendance to jot down the garments that were to their liking. Once selections were made, the client would sit with the designer, who would then fit the garments to that client's specific measurements and exact preferences, or a buyer would reproduce them for their own store.

Today, the couture collections are seen on the runways during Pairs Fashion Week. Pricing typically begins in the high thousands and can reach into the hundreds of thousands for these fine garments. Many companies use the glamour[30] and appeal of their couture collections, which account for a small market share of their overall business, as a catalyst[15] to boost sales for their ready-to-wear, accessories, and fragrance[31] business, which represent the bule of their revenue. Couture collections are often used as a "visual advertisement" to bring excitement to the brand and to elicit[16] sales for the more affordable ready-to-wear collection. Such as some well-known couture labels are Armani Privè, Atelier Versace, Chanel, Christian Dior, Givenchy, Jean Paul Gaultier, and Valentino.

The *Chambre Syndicale de la Haute Couture* accepts "foreign" members; however, there are only a handful of fashion designers outside of Paris who practice the fine technique of couture craftsmanship. Elie Saab, Giorgio Armani, and Paul Smith are examples. The French Ministry allows for outside members in an effort to show their strong belief in the importance of the globalization of the fashion industry. Ralph Rucci, Rick Owens, Adam Kimmel, Zac Posen, and Mainbocher are the only American designers to have achieved haute couture status. They have each been invited by the Ministry to show their collections in Paris and currently are, or have been, members of The *Chambre Syndicale de la Haute Couture*. Interestingly enough, Thom Browne, a New York-based menswear designer, independently showed his collection in Paris as a nonmember. A complete list of current members can be found at www.modeaparis.com.

通用词汇

[13] practitioner *n.* 从业者　　[15] catalyst *n.* 催化剂　　[16] elicit *vt.* 引出
[14] inherent *adj.* 固有的

高级时装屋由高级女装裁缝师监管一个车间里所有熟练工人从事的手工工艺，他们都是手工或裁剪方面的专家。根据设计师的偏好，这个过程也许从草图或效果图开始，或从立体裁剪和剪裁白坯布开始。为了完成一件高级时装，精致装饰、刺绣和修饰经常是从外面购买，有人专门从事这个行业，然后谨小慎微地缝合到服装上。精美合体是高级时装固有的品质。顾客要忍受很多次试衣，确定已经取得精确的尺寸，确保不仅精确地合身，而且还要有样式和舒适，它们同等重要。

当高级时装系列生产出来后，要首先在指定的展厅里以非公开形式展示给新闻媒体、买手和高端客户。每一个模特儿拿着一张与样式编号对应的卡片，使观众能够草草地记下他们喜欢的服装。客户将选中的款式告知设计师，设计师将按照客户的实际尺寸和准确参数提供相应的服装，或者买手将为自己的商店重新生产他们选中的款式。

今天，在巴黎时尚周上的 T 台上能够看到高级时装的发布会。这些上等的服装的价格最低几千，高达几万。很多公司利用高级时装的优雅和魅力作为催化剂，促进他们的成衣、配饰和香水生意的销售，因为这些代表了他们全年的收益率，而高级时装只占他们整个生意的一小部分。高级时装系列经常被用作视觉广告，给品牌带来活力，给相对便宜些的成衣带来更多的销售。例如一些知名的高级时装品牌：阿玛尼（Armani Privè）、范思哲（Atelier Versace）、夏奈尔（Chanel）、克里斯汀·迪奥（Christian Dior）、纪梵希（Givenchy）、让·保罗·戈尔捷（Jean Paul Gaultier）和华伦天奴（Valentino）。

巴黎高级时装协会接纳外国成员，但是巴黎以外只有很少的时装设计师从事高级时装的精致工艺技术，例如艾利·萨伯（Elie Saab）、乔治·阿玛尼（Giorgio Armani）和保罗·史密斯（Paul Smith）。法国政府部门允许外来成员在这样一个全球化时装行业占有重要地位的国度里施展他们的才能。拉尔夫·鲁奇（Ralph Rucci）、瑞克·欧文斯（Rick Owens）、亚当·基梅尔（Adam Kimmel）、扎克·珀森（Zac Posen）和梅因布彻（Mainbocher）是取得高级时装资格的美国设计师。他们被法国政府邀请到巴黎展示他们的新产品，或者他们已经是巴黎高级时装协会的成员。非常有趣的是，纽约的男装设计师桑姆·布郎尼（Thom Browne）没有协会成员资格，却在巴黎独立举行了他的发布会。www.modeaparis.com 网站上能够看到现时会员完整列表。

专业词汇

25. craft *n.* 技艺
26. muslin *n.* 平纹细布
27. toile *n.* 坯布
28. embellishment *n.* 修饰
29. trunk show 非公开时装表演
30. glamour *n.* 魅力
31. fragrance *n.* 香水

3. Bespoke

"Bespoke" is a British term used to describe individually crafted and patterned men's clothing. The Oxford English Dictionary defines "bespoke" as made-to-order clothing, made to each individual customer's precise measurements and specifications. Although bespoke is not a protected label, like couture. The Savile Row Bespoke Association (a professional organization consisting of Savile Row tailors) has attempted to set a standard by providing minimum requirements for a garment to be allowed the prestigious[17] use of its name. Savile Row is a very short street in central London, called "the golden mile of tailoring", famous for its bespoke tailors, among them Davies and Son, Gieves & Hawkes, and Norton and Sons. Historical Savile Row clients have included Napoleon III and Winston Churchill.

4. Designer

Also known as ready-to-wear (oftentimes abbreviated RTW) or "off-the-rack" and by the French term Prèt-à-porter, designer clothing is factory made and finished to fit standard sizes. Don't, however, let the phrase "off-the-rack" fool you. Whether mass produced or offered in limited quantities, designer clothing is exclusive and uses the finest imported fabrics and trims. Ready-to-wear collections are general presented twice a year (Spring/Summer and Autumn/Winter) during fashion weeks around the world, and they appear earlier than the couture collections. The price point can oftentimes exceed $1,000 per garment, but can range in lower price points or skyrocket to high three-figure numbers. Some of the most popular designer labels are Ralph Lauren, Donna Karan, Calvin Klein, Vera Wang, and Catherine Malandrino.

In Pairs, the *Chambre Syndicale du Prèt-à-Porter des Couturiers et des Crèateurs de Mode* is an established association, created in 1973, that is made up of all the fashion designers who produce ready-to-wear. The *Chambre Syndicale de la Mode Masculine* is an association that specifically includes the top menswear designers who produce ready-to-wear collections.

5. Bridge

Bridge garments are in between ready-to-wear and better, and carry a price point generally ranging in price from $300 to $600 per garment. Career wear and separates[32], along with dresses, are often indicative of a bridge classification. DKNY, CK, and Anne Klein II are examples of bridge labels.

6. Better

Better is one step down from bridge. Sportswear, various coordinates[18], separates, and dresses may all appear in better collection, and will typically sell for less than $600 per piece, but they primarily fall into a price point range of $150—$300. Some of the most popular better labels are Ellen Tracy, Henneth Cole, and Anne Klein.

3. 定制

"Bespoke"（定制）是英国用语，用来描述男性服装个性化的工艺和样式。牛津英语词典定义"bespoke"为根据每个客户的精确尺寸和详细规格定做的服装。尽管定制不是像高级时装有受保护的品牌，但是萨维尔街定制协会（由萨维尔街裁缝构成的专业组织）设法制定了一套标准，达到此标准最低要求的服装被允许有萨维尔街的荣誉使用权。萨维尔街在伦敦中心，是一条很短的街，被称之为"裁缝业的黄金一百米"，以定制的裁缝师而出名，他们当中有戴维斯父子（Davies&Son）、君皇仕/吉凡克斯（Gieves&Hawkes）和诺顿父子（Norton&Sons）。历史上，拿破仑三世和丘吉尔都是萨维尔街的客户。

4. 设计师品牌

也称为成衣（经常缩写为RTW）或者"现成的"，法语称Prêt-à-porter，设计师服装是根据标准尺寸由工厂制造完成。但是，你不要被短语"off-the-rack"所愚弄，不管是大规模或有限数量生产，设计师服装是高级的，使用最好的进口面料和装饰。成衣发布会通常在世界各地的时尚周期间举行，一年两次（春/夏和秋/冬），它们比高级时装发布会早。每件服装的价格时常超过1000美元，但范围可以从较低价位或扶摇直上到三位数。最有名的一些设计师品牌是拉尔夫·劳伦（Ralph Lauren）、唐娜·凯伦（Donna Karan）、卡尔文·克莱因（Calvin Klein）、王薇薇（Vera Wang）和凯瑟琳·玛兰蒂诺（Catherine Malandrino）。

在巴黎，"裁缝和时尚设计师"是一个建立于1973年的协会，由所有生产成衣的时尚设计师构成。"裁缝和时尚男性设计师"是专门生产男装成衣的设计师组织。

5. 过渡型

过渡型服装处于成衣和较好型服装之间，每件服装的价格带通常在300～600美元。职业服装、可配套穿的单件服装和连衣裙等通常归入此类。唐可娜儿（DKNY）、卡尔文·克莱因（CK）和安妮·克莱因Ⅱ（Anne Klein Ⅱ）就是过渡性服装品牌。

6. 较好型

较好型服装比过渡型服装低一等。运动型服装、各种可搭配穿的服装、单件服装、连衣裙，都可归入此类，每件服装通常不会超过600美元，价格带通常在150～300美元。最为人们熟知的"较好"品牌是爱伦·瑞丝（Ellen Tracy）、凯尼斯·柯尔（Henneth Cole）和安妮·克莱因（Anne Klein）。

专业词汇

32. separates *n.* 可搭配穿着的女服

通用词汇

⑰ prestigious *adj.* 受尊敬的　　⑱ coordinate *n.* 协调

7. Contemporary

Contemporary collections offer trendy apparel at a relatively affordable price point aimed at women in their twenties and thirties. Cynthia Steffe, Rebecca Taylor, and BCBGMAXAZRIA are all considered contemporary designers.

8. Secondary

Secondary lines are used by designers who want to offer a lower-priced line aside from their designer collection. The price points differ, but these fashions can generally be found for less than $300 per piece at retail. Marc by Marc Jacobs and Lauren by Ralph Lauren are considered secondary lines.

9. Moderate

Moderate fashions are promoted to the average, everyday customer and usually retail for less than $100 apiece. Some of the most popular moderate retailers are Liz Claiborne, Abercrombie & Fitch, Nine West, and the Gap.

10. Private label

Merchandise[33] that is manufactured by a store, or in partnership with an apparel manufacturer, is considered private label. Store advantages include greater control over production, cost[34], pricing, advertising budget[35], and design. Private label runs a gamut[19] of price points and is generally produced for the bridge to moderate markets. Some of the most successful private label businesses are International Concepts (I.N.C.) for Macy's and Hunt Club for J.C. Penney.

11. Mass

Mass market or budget caters to the lower end of the apparel continuum, with retail pricing generally under the $50 price point. Product categories[36] generally include casual sportswear such as T-shirts[37] and jeans. Some of the most popular budget retailers are OLD Navy, Target, Wal-Mart, Kmart, and Kohl's. Mass market is made in large quantities and is geared toward the general public.

12. Discount

Discount merchandise, also referred to as off-price, is excess merchandise that not sell at its full retail price through its original and intended retailer. These items can be found at varying price points in an array of retail outlets such as Filene's Basement (the inventor of the off-price store concept), Ross Stores, T.J. Maxx, Loehmann's, Marshalls, and Saks Fifth Avenue. Discount merchandise can also be found in factory outlet stores.

通用词汇

[19] gamut *n.* 全范围

7. 现代型

现代型服装是指潮流性服装，价格相对便宜，目标顾客为20岁到30岁的女性。辛西娅·史黛菲（Cynthia Steffe）、瑞贝卡·泰勒（Rebecca Taylor）和BCBGMAXAZRIA都被认为是现代型服装的设计师。

8. 二线品牌

二线品牌是指设计师除了他们的设计师品牌以外，还有一个价位低些的品牌。不同二线品牌价位不同，但通常每件零售价格不超过300美元。马克·雅可布之马克（Marc by Marc Jacobs）和拉尔夫·劳伦之劳伦（Lauren by Ralph Lauren）就是二线品牌。

9. 适中型

适中型是针对普通的日常消费者，通常每件价格不超过100美元。最受欢迎的适中型品牌有丽兹·克莱本（Liz Claiborne）、阿贝克隆比＆费奇（Abercrombie&Fitch）、玖熙（Nine West）和盖璞（Gap）。

10. 自有品牌

服装商品由服装商店或合作伙伴关系的服装生产商生产，这种品牌被认为是自有品牌。商店的优势是能够较好地控制产品、成本、定价、广告预算和设计。自有品牌价位变化性很大，通常从过渡类型服装到适中类型服装。最成功的自有品牌有美国顶级连锁百货公司梅西百货（Macy's）的International Concepts（I.N.C.）和J.C. Penney百货公司的Hunt Club。

11. 大众型

大众市场或价格低廉的市场，满足较低端的消费者。每件零售价格在50美元以下。产品种类通常包括休闲运动服装，例如T恤和牛仔。最熟知的大众市场零售商有老海军（Old Navy）、塔基特（Target）、沃尔玛（Wal-Mart）、凯马特（Kmart）和科尔士（Kohl's）。大众市场服装生产量大，针对普通大众。

12. 打折型

打折商品即降价商品，指按照零售商的正常定价未能卖出的剩余产品。这些商品在很多出清的零售商店都有销售，例如菲妮斯地下商场（Filene's Basement，降价商店概念的发明人）、罗斯服饰零售店（Ross Stores）、T.J. Maxx，Loehmann's，Marshalls和第五大道（Saks Fifth Avenue）。打折商品也能在工厂的出清商店里买到。

专业词汇

33. merchandise *n.* 商品
34. cost *n.* 成本
35. budget *n.* 预算
36. category *n.* 类型
37. T-shirt T恤

Within these price points, clothing classifications fall into various product categories, including women's, men's, young men's, collegiate, tweens (pre-teen), juniors, children's and layette[38] wear, formalwear[39], outwear, intimates[20], maternity, and swimwear.

在这些价位之内，服装有很多种类，包括女装、男装、年轻男装、学院装、青年装、少年装、童装、婴幼儿装、正装、外出装、内衣、孕妇装和泳装等。

专业词汇

38. layette *n.* 婴儿的全套服装   39. formalwear 正式服装

通用词汇

⑳ intimate *n.* 亲密

Passage 4

Fashion Design

For fashion designer, it is important to develop an awareness of your taste and style (not how you dress-designer are often the worst dressed in a room because they are too busy thinking about how to dress others). Not everyone has an aptitude[①] or desire to design "unconventional" clothes. Some designers focus on the understatements or detail of garments. Other designers design "conventional" garments, but it is the way they are put together that makes the original modern.

Knowing what you are best at is essential, but doesn't mean that you should not experiment. It can take a while to "know yourself" and this period of discovery is usually spent at college. There has to be a certain amount of soul searching; it's not so much being the designer that you want to be, but rather finding out the designer that you are. You must be true to your own vision of how you want to dress somebody.

Beyond that, the rest is in the hands of the industry and the fashion-buying public to decide, and for every person who likes your work there will be someone who doesn't. This is common and working in such a subjective[②] field can be confusing, but eventually you will learn to navigate your way through criticism and either develop a steely exterior[③] or recognize which opinions you respect and which to disregard[④]. Once you accept this, you are free to get on with what you are best at designing clothes.

1. Know your subject

If a career in fashion is what you want then you need to know your subject. This might appear to be an obvious statement, but it must be said. You may protest, "I don't want to be influenced by other designers' work". Of course not, but unless you know what has preceded[⑤] you, how do you know that you aren't naively reproducing someone else's work?

Magazines are a good place to start, but don't just automatically[⑥] reach for Elle and Vogue. There are many more magazines out there, each appealing to a different niche market and style subculture[⑦] and you should have a knowledge of as many as possible; they are all part of the fashion machine. Magazines will not only make you aware of different designers, but so-called lifestyle magazines will also make you aware of other design industries and culture events that often influence (or will be influenced by) fashion.

第 4 课

时装设计

对时装设计师来说，重要的是，要有发展自己品味和风格的意识（不是你穿着如何，在房间里设计师经常穿得最糟糕，因为他们总是忙于想着如何打扮别人）。不是所有人有才能或有欲望设计非传统服装。一些设计师专注于服装的内在或细节；一些设计师设计传统服装，也正是这样，他们共同制造了新颖的现代时尚。

知道你最擅长什么是重要的，但并不意味着你可以不去尝试其他。可以花一段时间认识自己，发现自我的这段时间通常是在大学，那是一种灵魂的反省。与其说你想成为一名设计师，不如说你想成为什么样的设计师。你必须真实地用自己的视角考虑如何去打扮别人。

除此而外，其他方面就是由行业和购买时装的大众决定，因为有人喜欢你的作品，也有人不喜欢你的作品。在这种主观性行业里工作，面临这样的困惑是很正常的。但是，通过批评最终你将学会驾驭你自己，或者锻炼自己钢铁般的内心，或者能够识别哪些意见值得你尊重，哪些意见值得放弃。一旦你接受这些，你便容易了解你最擅长设计什么服装。

1. 熟悉你的学科

如果你想要从事时装行业，你需要熟悉你的学科。这似乎是显而易见的，但必须要说明，你也许声称"我不想受到其他设计师作品的影响"。当然不是这样，除非你知道你的前人做了些什么，否则，你怎么知道你不是在天真地复制别人的作品？

杂志是起步的好地方，但不能想当然地就只看 *Elle* 和 *Vogue* 两本杂志。还有很多杂志，每一种杂志，关注不同的个性市场和亚文化风格，你必须尽可能地了解更多的知识，它们都是时尚机器的零部件。杂志不仅使你了解不同的设计师，同时，关于生活方式方面的杂志还将使你了解到其他的设计行业和影响时尚的文化事件（或受到时尚影响的文化事件）。

通用词汇

① aptitude *n.* 才能
② subjective *adj.* 主观的
③ exterior *n.* 外部
④ disregard *vt.* 不顾
⑤ precede *vt.* 先于
⑥ automatically *adv.* 自动地
⑦ subculture *n.* 亚文化

By regularly reading magazines you will also become aware of stylists, journalists, fashion photographers and hair and make-up artists, models, muses¹, brands and shops that are all-important to the success of a fashion designer.

There are also some great websites that show images of outfits on the catwalk almost as soon as the show has taken place.

2. Starting your research

Designers are like magpies⑧, always on the lookout for something to use or steal! Fashion moves incredibly⑨ fast compared to other creative industries and it can feel like there is constant pressure to reinvent the wheel each season. Designers need to be continually seeking new inspiration in order to keep their work fresh, contemporary, and above all, to keep themselves stimulated⑩.

In this sense, research means creative investigation, and good design can't happen without some form of research. It feeds the imagination and inspires the creative mind.

Research takes two forms. The first kind is sourcing material and practical elements. Many fledgling⑪ designers forget that finding fabrics and other ingredients⑫—rivets², fastenings³ or fabric treatments⁴, for example—must make up part of the process of research and having an appreciation⑬ of what is available, where from, and for how much, is essential.

The second form of research is the kind you make once you've found a theme or concept for use in your designs. Themes can be personal, abstract or more literal. Alexander McQueen, Vivienne Westwood and John Galliano have designed collections where the sources of inspiration are clear for anyone to see. McQueen's "It's A Jungle Out There" 1997—1998 collection mixed religious paining with the evocation⑭ of an endangered⑮ African antelope. Westwood has drawn on pirates, the paintings of Fragonard and 17th- and 18th-century decorative arts in the Wallace Collection for inspiration in different collections. Galliano has been influenced by the circus, ancient Egypt, punk singer Siouxsie Sioux and the French Revolution.

Designers may also convey a mood⁵ or use a muse for inspiration. Galliano currently cites singer Gwen Stefani as a muse, but has also based collections around 1920s' dancer Josephine Baker and Napoleon's Empress Josephine.

Using a theme or concept makes sense because it will hold together the body of work, giving it continuity and coherence⁶. It also sets certain boundaries⁷—which of course the designer is free to break—but having a theme initially⑯ gives the designer focus.

专业词汇

1. muse *n.* 缪斯
2. rivet *n.* 铆钉
3. fastening *n.* 系扣
4. fabric treatment 面料后处理
5. mood *n.* 情绪
6. coherence *n.* 统一
7. boundary *n.* 边线

通过定期阅读杂志，你将对造型师、新闻记者、时尚摄影师、发型师、化妆师、模特儿、缪斯、品牌和商店有更多的了解，所有这些对于时尚设计师的成功都是很重要的。

另外，还有很多网站张贴了 T 台上展示的服装，它们发布的速度几乎与 T 台同步。

2. 开始你的研究

设计师就像喜鹊那样，总是蹲守在瞭望台上，寻找可以使用或"偷"的东西。与其他设计行业相比，时装的变化速度令人难以置信，你感觉到持续不断的压力，每一个季节都要改变新的形式。为了保持作品的新鲜度和时代感，设计师需要持续寻找新的灵感。总之，要使自己处于兴奋状态。

在这种意义上来说，研究意味着创造性调查，如果没有某种形式的研究，就没有好的设计产生。调查促进想象，激发创造性的智慧。

研究有两种形式。第一种寻找材料和可得到原料的来源。很多没有经验的设计师往往忽视寻找面料和其他辅料，例如：铆钉、松紧带、面料的处理，它们必须是研究过程中的部分工作。能够评估什么是可以得到的，它们来自哪里，需要多少是十分重要的。

第二种研究形式是在你的设计中寻找一个主题或概念。主题可以是个性的、抽象的、或比较字面化的。亚历山大·麦昆（Alexander McQueen）、维维安·韦斯特伍德（Vivienne Westwood）和约翰·加利亚诺（John Galliano）设计的作品，任何人都可以清楚地看到灵感源。麦昆 1997—1998 年的"丛林之中"系列融合了宗教绘画以及对非洲羚羊濒临灭绝的呼唤。韦斯特伍德在不同系列中，从弗拉戈纳尔（Fragonard）的"海盗"绘画和华莱士收藏馆（Wallace Collection）里 17、18 世纪的装饰艺术吸收灵感。加利亚诺（Galliano）受到了马戏、古埃及、朋克歌手苏克西（Siouxsie Sioux）和法国大革命的影响。

设计师还可以传达一种情绪，或者以一位缪斯为灵感。加利亚诺以歌手格温·史蒂芬妮（Gwen Stefani）作为他的缪斯，在他的系列中还以 1920 年代的舞蹈家约瑟芬·贝克（Josephine Baker）和拿破仑皇后约瑟芬（Josephine）为灵感。

使用一个主题或概念将会找到一种感觉，能够把握作品的主体，使其具有连贯性和一致性。它也会设定某些边界，当然，设计师可以任意地打破，但是，有了主题设计师可以更加集中注意力。

通用词汇

⑧ magpie *n.* 喜鹊
⑨ incredibly *adv.* 不可思议的
⑩ stimulate *vt.* 刺激
⑪ fledgling *adj.* 刚开始的
⑫ ingredient *n.* 组成部分
⑬ appreciation *n.* 欣赏
⑭ evocation *n.* 唤出
⑮ endanger *vt.* 使遭受危险
⑯ initially *adv.* 开始

3. Choosing a concept

When choosing a theme, be honest, it needs to be something that you can work and live with for the duration[17] of the collection. This means that it should be a subject that you are interested in, that stimulates you and that you understand.

Some designers prefer to work with an abstract[8] concept that they want to express through the clothing (for example, "isolation"), while others want to use something more visually orientated (such as "the circus").

Either of these approaches is appropriate and it is about choosing which works for you. But it does need to work for you; it is pointless choosing a theme that doesn't inspire you. If the ideas are still struggling to come after a certain point a clever designer will be honest and question their choice of theme.

Remember, press and buyers are generally only interested in the outcome. Do the clothes look good? Do they flatter? Do they excite? Will they sell? They are not necessarily interested how well you've managed to express quantum[18] physics through a jacket. But if this is what you want to express, then do it.

4. Research and the sources

Where to go begin your research depends on your theme or concept. For an enquiring designer the act of researching is like detective work, hunting down elusive information and subject material that will ignite[19] a spark.

The easiest place to start to research is on the internet. The Web is a fantastic source of images and information. It is also greet for sourcing fabrics direct from manufactures that produce socialist material or companies that perform specific services.

A good library is a treasure. Libraries are geared to provide books to a broad cross-section of the community so tend to have a few books about many subjects. Specialist libraries are the most rewarding, and the older the library the better-books that are long out of print will(hopefully) still be on the shelves, or at least viewable upon request. Colleges and universities should have a library geared towards the courses that are being taught, though access may be restricted if you are not actually studying there.

Flea markets and antique[20] fair are useful sources of inspirational objects and materials for designers. It goes without saying that clothing of any kind, be it antique or contemporary, can inspire more clothes. Historic, ethnic[21] or specialist clothing-military garments, for example, offer insight into details, methods of manufacture and construction that you may not have encountered before.

Like flea markets, charity shops are great places to find clothes, books, records and bric-a-brac that, in the right hands and with a little imagination, could prove inspirational. Everyday objects that are no longer popular or are perceived as kitsch[22] can be appropriated, rediscovered and used ironically to design clothes.

3. 选择一个概念

诚实地选择一个主题，此系列持续的时间内，要使你能为之工作，并与之共处。也就是说你感兴趣的、你为之激动和了解的主题。

一些设计师喜欢通过服装来表达抽象概念（例如隔离），另外的设计师想使用更形象的概念（例如圆圈）。

这两种方法都是可行的，就看哪种选择适合你。确实需要适合你的主题，选择一个不能使你兴奋的主题是毫无意义的。如果一段时间过后，理念仍然姗姗来迟，聪明的设计师就要坦诚地质疑所选择的主题。

记住，媒体和买手一般只对结果感兴趣。那些衣服看上去漂亮吗？它们使人显得更漂亮吗？它们令人激动吗？它们会有销售吗？他们没有必要关心你如何设法在一件夹克上表达量子物理学，但是如果这是你想表达的，那你就去做吧。

4. 研究和资源

从哪里开始研究，这要根据你的主题或概念而定。对于好学的设计师来说，研究就像是侦探工作，去挖掘将会燃起火花的独特信息和主题材料。

开始研究最便利的地方是互联网。网络是图片和信息的神奇资源库。从网络也能直接了解面料制造商的信息，它们可能生产社会化大众面料，也有一些公司能够提供特别的服务。

一个好的图书馆也值得珍惜。图书馆提供的书籍往往考虑到社会不同行业的需求，因此，只有少量的专业书籍。专业图书馆最有收获，越是老的图书馆越好，往往有一些绝版的书籍仍然陈列在书架上，或至少根据要求能看到。学院和大学应该有一个根据所教专业配套相应图书的图书馆。但是如果你不在那里学习，进入图书馆将受到限制。

跳蚤市场和古玩市场也是十分有用的资源库，设计师可以从那里找到启发灵感的物品和资料。毋庸置疑，那里有各种各样的服装，古代的或当代的，能够带来更多灵感设计。例如，历史的、民族的、部队的制服，使你看到以前从未见过的细节、制造方法和结构。

和跳蚤市场一样，慈善商店也是寻找服装、书籍、录像和小古董的好地方。正确使用，并稍加一点想象，就能显现为灵感。每天都有一些不再受欢迎或者被认为是劣质的东西，被人们重新欣赏、再次发现和嘲讽地使用它们去设计服装。

专业词汇

8. abstract *adj.* 抽象

通用词汇

⑰ duration *n.* 持续
⑱ quantum *n.* 量子
⑲ ignite *vt.* 点燃
⑳ antique *adj.* 古老的
㉑ ethnic *adj.* 种族的
㉒ kitsch *n.* 粗劣的作品

Museums, such as London's Victoria and Albert Museum, not only collect and showcase interesting objects from around the world, both historical and contemporary, but also have an excellent collection of costume that can be viewed upon request.

Large companies, with the budget, send their designers on research trips, often abroad, to search for inspiration. There, the designers are armed with a research budget and a camera, and can record and buy anything that might prove useful for the coming season. Designers with a tight budget might use a holiday abroad as a similar opportunity.

Sources of images can be photocopies, postcards, photographs, tear sheets from magazines and drawing. But anything can be used for research: images, fabric, details such us buttons or an antique collar[9]—anything that inspires you qualifies as research. Whichever items you collect must be within easy reach (and view) so that you have your reference constantly about you.

5. The research book

As a designer you will eventually develop an individual approach to "processing" this research. Some designers collect piles of photocopies and fabrics that may find their way on to a wall in the studio. Others compile research or sketchbooks where images, fabrics and trimmings are collected and collated, recording the origin and evolution of a collection. Still others take the essence of the research and produce what are called mood, theme or storyboards.

A research book is not necessarily solely for the designer's use. Showing research to other people is useful when trying to convey the themes of a collection. It might be used to communicate your concept to your tutor, your employers, employees or a stylist.

Research books are not just scrapbooks. A scrapbook infers that the information is collected, but unprocessed. There is nothing duller than looking through pages of lifeless, rectangular[10] images that have been (too) carefully cut out. It is also debatable[23] how much the designer has gleaned from creating pages like this. A research book should reflect the thought processes and personal approach to the project. It becomes more personal when it is drawn on and written in, and when the images and materials that have been collected are manipulated[24] or collaged[11].

6. Collage

The word "collage" is derived from French word for glue. A good collage is where the separate elements (images) work on different levels at the same time, to form both a whole and also its individual component parts.

Successful collage usually includes a bricolage[12] of different-sized, differently sourced images that provide a stimulating visual rhythm.

7. Drawing

Drawing a part or the whole of a picture you have collected as research helps you to understand

博物馆，例如伦敦维多利亚和艾尔伯特博物馆，不仅收藏和展示来自世界各地历史的和当代的奇珍异品，而且还收藏了精湛的古代服装，提出申请就可以看到它们。

一些有预算的大公司提供设计师研究性的旅行，到国外去寻找灵感。在那里，设计师带着研究经费和一台相机，记录和购买对下一季有用的东西。预算紧张的设计师也可以利用去国外度假，当作类似的机会。

图像资源可以是照片复印件、明信片、照片、从杂志和绘画上撕下的小碎片。但是可以从任何东西中进行研究：图像、面料、纽扣或古代的领子等细节，任何能够激起灵感的东西都可以用来研究。不管你收集什么，必须是在能够得到（和看到）的范围内，这样在你周围你有源源不断的参考资料。

5. 研究本

作为一个设计师，最终你要形成一种个性化研究过程的方法。一些设计师也许找到了他们的方法，就是将收集的一大堆照片复印件和面料贴在工作室的墙上；还有一些设计师编辑所做的研究或草稿本，即将图片、面料和装饰品放在一起比对，记录一个系列的起源和发展过程。还有一些设计师汲取研究的精华，根据情绪、主题或故事板进行编辑。

研究本不仅仅对设计师有用，当要告知别人系列的主题，用研究本传达信息显得尤其有效。你可以借助研究本将你的概念告诉你的导师、你的雇员、你的雇主或造型师。

研究本不是剪贴簿。一本剪贴簿意味着是信息的收集，没有发展进程。看着那些没有生气的、仔细剪下的长方形图片是很乏味的事情。有多少设计师这样做是值得商榷的。研究本应该反映从事这个项目的思想过程和个性化方法。如果在研究本上有你的绘画、你写的文字，并将收集的图片和资料拼贴在一起时，研究本就更加个性化。

6. 拼贴

"拼贴"来源于法语"黏贴"一词。好的拼贴是同时将不同层面的各个元素和图形放在一起，形成一个整体，同时个体又各具个性。

成功的拼贴通常包括不同尺寸、不同来源的图片拼贴在一起，产生令人兴奋的视觉节奏。

7. 绘画

在研究时，从你收集的某幅图片中画它的局部或全部，将帮助你理解构图的形状和形式，

专业词汇

9. collar *n.* 衣领
10. rectangular *adj.* 长方形的
11. collage *vt.* 拼贴
12. bricolage *n.* 拼装

通用词汇

㉓ debatable *adj.* 可争辩的
㉔ manipulate *vt.* 熟练控制

the shapes and forms that make up the image, which, in turn, enables you to appreciate and utilize the same curve in a design or when cutting a pattern.

Using collage and making your own drawings allows you to deconstruct[25] an image such as a photograph, photocopy, drawing or postcard. This is necessary because it may not be the whole image that will ultimately be useful to your designs; a picture may have been chosen for its "whole", but it is only when it has been examined in more depth[13] that other useful elements may be discovered. For example, a photograph of a Gothic cathedral is rich in decorative[14] flourishes[15], but it almost needs a magnifying glass to be able to understand the detail. By cutting up an image[16] or using a "viewfinder"[26]— a rectangle paper "frame" that enables you to focus on part of an image, much like the viewfinder on a camera-smaller elements or details can become more apparent and be more easily examined.

8. Juxtaposition[27]

Placing images and fabrics together on the pages of your research book will help you to make important decisions about the content of your designs. Sometimes disparate images or materials may share similarities even though they are essentially different. For example, the spiral[17] shape of an ammonite fossil[28] is similar to a spiral staircase or a rosette[18]. Or an image may be suggestive of a fabric you have sourced, for example, a place of velvet[19] may evoke the texture of moss and lead you to think about natural imagery.

By utilizing drawing, collage and juxtaposition in your research books, you are processing and analyzing what has been collected. You are able to render and interpret images and materials as part of your own logical progression or journey.

9. Mood, theme and storyboards

Mood, theme and storyboards are essentially a distillation[29] of research. In a sense they are the "presentation" version of the research book. They are made up as collage, and as the name suggests, generally mounted on board, which makes them more durable. They are used by a designer to communicate the themes, concepts, colors and fabrics that will be used to design the season's collection. They may include key words that convey a "feeling", such as "comfort[20]" or "seduction[21]". If the collection must be tailored to a particular client, the images may be more specifically attuned to the perceived lifestyle/identity of the potential client.

10. Designing

Once your research has been collated[30], you can start on design. But there is nothing more intimidating than a blank page. The process can be very frustrating; even when the designs start to come it can take a while before any of them are very satisfactory. This is a natural part of the design process. Many early designs are thrown away and you might even begin to question your abilities. Don't panic! It takes time to hit your stride, and after sweating a while over the page better ideas will

反过来，在你设计或裁剪图形时，使你能够领会和利用相同的曲线。

使用拼贴和自己绘图使你能够解构一幅图形，例如照片、复印图片、绘画和明信片。这样做是必要的，因为不是整幅图片对你的设计有用。一幅图片也许整幅都被选来，但是，仔细研究之后，可能会发现其中的一部分元素更有用。例如，一幅反映了辉煌装饰的哥特式教堂图片，几乎需要用放大镜才能理解它的细节。但是剪开一幅图片，或者使用"取景器"——一种长方形框子的纸片，很像照相机的取景器，使你能够集中图片的局部，很小的元素或细节变得更加明显，更加容易审视。

8. 并置

将图片和面料放在你研究本不同的页面上，将帮助你做出关于设计内容的重要决定。有时，在本质上完全不同的图片或资料也许会有共同特点。例如，一个菊石化石的螺旋形状与楼梯、或玫瑰花相似。或者一幅图片与你寻找的一块面料相似。例如，一块天鹅绒面料想起苔藓的肌理，使你想起自然的景象。

在你的研究本上使用绘画、拼贴和并置，处理和分析已经收集的资料，你将能够提供图片和资料的解释，使它们成为你逻辑进程的一部分。

9. 情趣、主题和故事版

情趣、主题和故事版从本质上是研究的升华。实际上，是研究本的呈现版本。顾名思义，由于由拼贴构成，通常放置在一张页面上，使它们更具有持久性。设计师用它交流主题、概念、色彩和面料，这些将被用来设计季节的系列产品。它们可能包括一些传达某种"情感"的关键词，例如舒适或者诱惑。如果这个系列是为某个特定的客户制作，图形将更加特别，要与潜在客户感知的生活方式和身份相协调。

10. 设计

一旦研究完成，你可以开始设计。但是，没有什么比一张白纸更恐惧。这个过程可能是相当令人沮丧的，甚至设计开始以后的一段时间内，没有一个设计是满意的，这是设计过程中的正常部分。开始时扔掉很多设计，你甚至开始怀疑你的能力。不要惊慌！迈开步子需要时间，面对纸张出一会儿汗以后，更好的理念才会出现。在这个阶段探索每一种可能，不要

专业词汇

13. depth *n.* 深度
14. decorative *adj.* 装饰
15. flourish *n.* 花样
16. image *n.* 影像
17. spiral *n.* 螺旋线
18. rosette *n.* 玫瑰花结
19. velvet *n.* 丝绒
20. comfort *n.* 舒适
21. seduction *n.* 诱惑

通用词汇

㉕ deconstruct *vt.* 解构
㉖ viewfinder *n.* 取景器
㉗ juxtaposition *n.* 并置
㉘ fossil *n.* 化石
㉙ distillation *n.* 蒸馏
㉚ collate *vt.* 校对

start to emerge. Explore every possibility that comes to mind and discard nothing at this stage. You might discover the potential of an ideal later on when you look back over your designs.

A designer's identity or style comes with time, but as well as that, the clothed themselves need an identity or to form part of a vision in order to stand apart from the competition. Certain elements should run through the designs to give them coherence. It could be where an armhole[22] is cut, the placement of a seam[23] on the body in a particular way, or a method of finishing[24] the fabric. If these elements tie in strongly with your theme to work as a "whole" you are on your way to making a real statement with your designs.

放弃任何东西。当你回头看你的设计时,你可能发现某种具有潜力的理念。

设计师的特征或风格形成需假以时日,为了在竞争中脱颖而出,设计师需要有特征,或者将特征作为视觉构成的一部分。某些元素应该贯穿设计,使它们取得和谐。可能在袖窿处裁剪,用某种方式在身体某个部位放置一条缝线,或者面料的一种后处理方法。如果这些元素与你的主题紧密相连、构成整体,你的设计就真正体现出你的独到之处。

专业词汇

22. armhole *n.* 袖窿   23. seam *n.* 接缝   24. finishing *adj.* 修整

Passage 5

Color

Color grabs[①] customers' attention, makes an emotional connection, and leads them to the product. Even when the basic product stays the same, changing the color gives a sense of something new.

The human eye can discern[②] 350,000 colors, but the human memory for color is poor. Most people cannot remember a specific color for more than a few seconds. In everyday conversations about color, it is sufficient to refer to a few colors by name—red, yellow, green, blue, white, and black—and add a qualifier—light[1] or dark[2], bright[3] or dull[4], cool[5] or warm[6]. General terms are not sufficient for communicating color information for design and manufacturing. Exact identification, matching[7], and reproduction of colors require an effective system with colors arranged in sequential order and identified with numbers and letters. Systems such as this are based on the basic three characteristics of color—hue[8], saturation[9], and value[10] (Figure 6.1).

Hue refers to the color—each color system designates a set of basic colors. Varying the other two characteristics[11] fills out the system.

Saturation (also called intensity[12] or chroma[13]) refers to the strength or purity[14] of the color and value to the lightness or darkness of the color.

The term tint applies to any color when white is added. Shade[15] refers to colors mixed with black. Tone[16] describes a grayed color. Creating a new tint, shade, or tone does not change the designation of the hue, but it does change the value and intensity of a color.

Color systems provide notation systems for the reproduction of colors and guidelines for harmonious[17] color groupings. The color wheel (Figure 6.2) is the simplest version of such a system. The primary colors are yellow, red, and blue; secondary colors are mixed[18] from two primaries (yellow + blue = green, yellow + red = orange[19], red + blue = violet[20]). Tertiary colors are mixed from one primary color and one secondary color (yellow + green = yellow-green). Relationships on the color wheel help designers select coordinate colors. When a patterned fabric or an ensemble uses colors next to each other on the color wheel, the color scheme is called analogous—for example, yellow, yellow green, and green.

通用词汇

① grab *vt.* 夺取　　　　　　② discern *vt.* 辨明

第 5 课

色 彩

  色彩能够吸引客户的注意，产生与产品相关联的情感。甚至当基本产品保持相同时，如果改变色彩，会产生新的感觉。

  人的眼睛能够辨别 350 000 种色彩，但是，人的色彩记忆很差。大多数人记得某种特别的色彩不超过几秒钟。在日常生活中讨论的色彩，几种就足够了，它们的名字是红、黄、绿、蓝、白和黑，再加一些修饰词——浅和深、明和暗、冷和暖。在设计和制造时，一般的术语不够用来交流色彩信息。准确地识别、匹配和生产色彩需要有效的体系，将色彩按照一定的序列进行排列，根据数字和字母进行识别。这种体系基于色彩的三个基本特征：色相、饱和度和明度（见图 6.1）。

  色相就是颜色的名称。每个色彩体系指定了一组基本的色彩。变化色彩的其他两个特征就构成完整性的色彩体系。

  饱和度（也称为密度或浓度），是指色彩的力度或纯度。而明度是指色彩的淡或深。

  浅色是指任何添加了白色以后的色彩。暗色是指混合黑色后的色彩。色调用来描述加灰的色彩。创造新的浅色、暗色和色调不改变色相，但是改变了色彩的明暗和浓度。

  色彩体系是一种标记体系，用于色彩的再生产，对取得和谐的色彩群组有指导作用。色轮（见图 6.2）是一种最简单形式的体系。原色是黄、红和蓝；间色是两种原色的混合（黄+蓝=绿，黄+红=橙，红+蓝=紫）。第三色是一种原色和一种间色的混合（黄+绿=黄绿）。色轮上的色彩关系有助于设计师选择调和的色彩。面料图案或色彩组合时，使用色轮上彼此靠近的色彩，这种色彩方案叫做同类色，例如，黄、黄绿和绿。

专业词汇

1. light *adj.* 淡色
2. dark *adj.* 深色
3. bright *adj.* 亮色
4. dull *adj.* 暗色
5. cool *adj.* 冷色
6. warm *adj.* 暖色
7. match *vi.* 匹配
8. hue *n.* 色相
9. saturation *n.* 饱和度
10. value *n.* 明暗
11. characteristic *adj.* 特征
12. intensity *n.* 纯度
13. chroma *n.* 色度
14. purity *n.* 纯度
15. shade *n.* 阴影
16. tone *n.* 色调
17. harmonious *adj.* 和谐的
18. mix *vt.* 混合
19. orange *n.* 橙色
20. violet *n.* 紫罗兰

Complementary[3] colors—colors opposite each other on the color wheel (also known as color complements)—intensify each other when used in combination even when they are mixed with white, black, or gray. The strong relationship between complement leads to variations: the double complement (two sets of complementary colors) and the split complements (a color plus the two colors on either side of its complement).

Three colors spaced equidistant on the color wheel are called a triad[4]—red, yellow, and blue make the primary triad; Green, orange, and violet make the secondary triad. An infinite number of color combinations can be developed by varying the value and intensity of the colors.

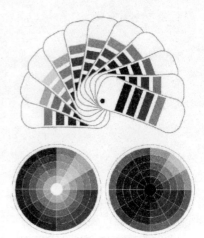

Figure 6.1 The pantone professional color system
图 6.1 潘通专业色彩体系

There are two kinds of color systems in American:

• The Munsell Color System, which includes a color atlas, the Munsell Book of Color (1976), with about 1,600 chips arranged in equal steps of hue, value, and chroma (intensity or saturation) and a notation for each.

• The Pantone® Professional Color System, which includes a color atlas, The Pantone Book of Color (1990), with 1,225 colors identified by name and color code. A color notation system may look incomprehensible at first glance, but to color professionals the system becomes a precise language for color identification.

The complete Munsell notation is written symbolically[21]: H(ue) V(alue)/C(hroma). For a vivid red, the notation would read 5R 6/14. For finer definitions, decimals are used—5.3R 6.1/14.4. Beginning with 1700 hues, the Pantone company expanded the range by 56 colors in 1998 and by 175 colors in 2002 in an attempt to provide a comprehensive mix of colors for fashion, architecture and interior design, and industrial design.

1. Color cycles

Color cycles refer to two phenomena: the periodic shifts in color preferences and the patterns of repetition[22] in the popularity of colors. Both depend on the mechanism[23] of boredom—people get tired of what they have and seek something new. New colors are introduced to the marketplace[24], available to consumers in product categories from fashion to interiors to automobiles. There is a lag time between the introduction of a new color or new color direction and its acceptance while people gain familiarity with the idea. Colors and color palettes[25] move from trendy to mainstream[26]. In time, interest in the colors wane, and they are replaced by the next new thing. This mechanism means that colors have somewhat predictable life cycles. It also means that colors that were once popular can be repositioned in a future season—the harvest gold of the 1970s became the sunflower gold of the 1990s.

补色——在色轮上相对的色彩（也称为色彩的补色）——当它们组合在一起时浓度要相等，甚至在混合白、黑或灰色时也要使浓度一致。补色之间的强度关系产生变量：双互补色（两个互补的色彩）和分裂互补（一种色彩与它补色两边的色彩构成的互补）。

在色轮上构成等边的三种色彩称为三色组合——红、黄和蓝构成三原色组合；绿、橙和紫色构成间色的三色组合。通过改变色彩的纯度和浓度可以得到无穷的色彩组合。

在美国有两种色彩体系：

• 蒙赛尔色彩体系，*Munsell Book of Color*（1976）书中有 1 600 块色彩图标集，它们按照色相、纯度和浓度（密度和饱和度）等距离地排列，每一种色彩都有标注。

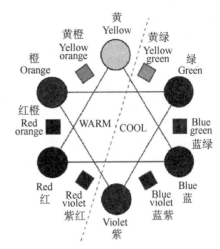

Figure 6.2　Color systems provide guidelines for harmonious color groupings

图 6.2　色彩体系提供了和谐的色彩群组指导

• 潘通专业色彩体系，包含了色彩图集。*The Pantone Book of Color*（1990）书中有 1 225 块色彩图标集，由名称和色彩代码来识别。乍一看，色彩的符号体系看上去十分难懂，但是对于色彩专业人士来说，这种体系成了识别色彩的精确语言。

完整的蒙赛尔标注符号为：H（ue）V（alue）/C（hroma）。例如，鲜艳的红色表示为 5R 6/14。要想更加精确的定义，可以使用小数点 5.3R 6.1/14.4。潘通公司于 1998 年在起初的 1 700 种色彩上增加了 56 种，2002 年又增加了 175 种。目的是努力为时尚、建筑和室内设计提供广泛的色彩调配。

1. 色彩周期

色彩周期是指两种现象：色彩偏爱的周期性变化和色彩流行模式的反复。两者都是因为厌倦心理所致——人们对拥有的东西感到疲倦，寻求新的东西。新的色彩被引入市场，然后到达消费者，体现在时尚、室内设计、汽车等诸多产品上。在新色彩或新色彩趋势引入后，到人们熟悉这种理念并接受它，有一段时间的滞后性。色彩和主色调从潮流进入到主流，同时，当一种色彩的兴趣减退时，就有另一种新的色彩替代它，这种机制说明色彩的生命周期可以预测，且曾经流行的色彩在未来季节里将再次流行。1970 年代丰收金色在 1990 年代变成了向日葵金色。

专业词汇

21. symbolical *adj.* 象征的　　23. mechanism *n.* 机制　　25. palette *n.* 调色板
22. repetition *n.* 反复　　　　24. marketplace *n.* 市场　　26. mainstream *n.* 主流

通用词汇

③ complementary *adj.* 互补的　　④ triad *n.* 三个一组

The vogue for a group of colors evolves over a period of 10 to 12 years, reaching its peak in mid-cycle. The color usually appears in fashion, moving quickly to high-end interiors, and then to the rest of the consumer products market.

2. Long-wave cycles

The recurrences of color themes can be traced not just in decades but in centuries. In the Victorian era, Owen Jones chose rich, bright, primary colors for London's Crystal Palace in 1851. He defended his choices saying that these same colors had been used in the architecture of the ancient Greeks. To some color historians, this comment illustrates the cycling of color through history, specifically the periodic return to primary hues. The cycle begins with bright, saturated, primary colors; this is followed by an exploration of mixed, less intense colors; then, it pauses in neutral[27] until the rich, strong colors are rediscovered.

Researchers have confirmed a periodic swing (Figure 6.3) from high chroma colors, to "multicoloredness," to subdued[28] colors, to earth tones, to achromatic[5] colors (black, white, and gray), and back to high chroma colors. In the period between 1860 and the 1990s, there were four marked color cycles lasting between 15 and 25 years.

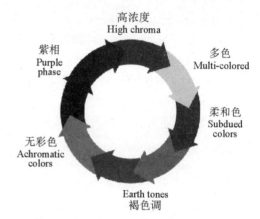

Figure 6.3　A periodic swing of colors
图 6.3　色彩周期循环

3. Color cycles and cultural shifts

Color cycles can be sparked by new technology. This happened about ten years after the opening of the Crystal Palace in the mid-1800s. One of the first synthetic[29] dyes[30] was introduced by the French: a color-fast purple[31] called mauve[32]. The color became the rage[33]—Queen Victoria wore the color to the International Exhibition of 1862—and gave its name to the Mauve Decade. Other strong synthetic dyes for red and green soon followed, allowing for a strong color story in women's clothing. Something similar happened in the 1950s when the first affordable[34] cotton[35] reactive dye for turquoise[36] led to a fad for the color and moved it from eveningwear into sportswear.

Economic conditions also disturb color cycles and start new ones. In the depressed 1930s, fabrics and colors were chosen as investments—the buffs[37], grays, and subdued greens and blues were low-key, could be worn more often, didn't show dirt, thus seldom need to be cleaned and, therefore, lasted longer. The steep drop in the stock market in 1987 coincided with the eclipse of bright color as the fashion look of the 1980s and the ascendance of Japanese design featuring austere[38], minimalist[39] black clothing.

一组色彩演变的时尚周期为 10～12 年，在周期中间阶段达到高峰。色彩通常在时装中出现，很快传播到高端室内用品，然后到达其他消费产品市场。

2. 长波周期

色彩主题的回归可能追溯到几个年代以前，甚至几个世纪以前。维多利亚时代，欧文·琼斯（Owen Jones）在 1851 年为伦敦的水晶宫选择了丰富亮丽的原色色彩。他为自己的选择争辩到，这些色彩与古希腊建筑中的色彩相同。对于一些色彩历史学家来说，这种评论注释了色彩的循环具有历史性，特别是周期性地复苏原色。周期开始于明亮、饱和的原色，然后就探索混合色，减少色彩的浓度，接下来是中性色，直到丰富、浓烈的色彩再次出现。

研究者已经证实色彩的周期性回转（见图 6.3）。从高浓度色彩、到多彩色、柔和色、褐色调、到无彩色（黑、白和灰），再回到高浓度色彩。从 1860 年到 1990 年代这段时间，色彩有四次显著的循环，每次循环大约持续 15～25 年。

3. 色彩周期和文化变迁

新技术引发色彩变化，这大约发生在 19 世纪中期水晶宫开幕后十年。法国人发明了一种活性紫色的合成染料，称之为苯胺紫。这种色彩疯狂流行，被称为苯胺紫年代。1862 年，维多利亚女王穿着这种色彩的服装去参观国际博览会。很快出现了其他浓烈的合成染料，如红色、绿色，使女性服装大放异彩。1950 年代也发生了类似的情形，首次出现了价格便宜的适合棉织物染色的绿松石染料，掀起了一股热潮，从晚装到运动服都采用这种色彩。

经济条件也会扰乱色彩周期并产生新的色彩。在 1930 年代经济萧条时，面料和色彩的选择像投资一样很低调——黄褐色、灰色和暗绿色以及蓝色，经常穿不显脏，因此洗涤次数减少，也能持续很长时间。1987 年急剧下跌的股市应和了 1980 年代时尚面貌中亮丽色彩的褪却，也应和了日本简朴设计风格和极简主义黑色服装的兴起。

专业词汇

27. neutral *adj.* 中性色
28. subdued *adj.* 柔和色
29. synthetic *adj.* 合成
30. dye *n.* 染料
31. purple *n.* 紫色
32. mauve *n.* 淡紫色
33. rage *n.* 风靡一时
34. affordable *adj.* 买得起
35. cotton *n.* 棉花
36. turquoise *n.* 蓝绿色
37. buff *n.* 米色
38. austere *adj.* 朴素的
39. minimalist *adj.* 极简主义

通用词汇

⑤ achromatic *adj.* 消色差的

Color cycles can be associated with social change. A visible cycle was identified by June Roche, a corporate color analyst, in the mid-1980s—the shift between colors associated with femininity and those influenced by men's fashion. She characterized the end of the 1970s as a dark phase in men's clothing with dusty[40] colors that were called elegant[41], refined[42], and sophisticated[43]. Use of these typical grayed European colors was new for American women's fashion but coincided with women's entry into fields such as finance, law, and medicine—fields formerly dominated by men. By the mid-1980s, those grayed colors looked dirty, and there was a shift to feminine colors. The shift between ultra-femininity and a suggestion of masculinity has been part of the fashion scene since Coco Chanel first introduced women to the concept of borrowing from men's closets[44] in the 1920s.

色彩周期与社会变化相关。一位企业色彩分析师 June Roche 总结出一个明显的色彩周期，1980 年代中期，色彩在女性化和男性化之间变化。她指出，1970 年代后期，男性服装呈尘土色的深色区间，被称为优雅、精致和成熟。尽管美国女性时装使用这些典型的欧洲灰色调是新潮，但是它与女性进入一些领域，例如金融、法律和医学相一致，这些领域过去由男性主宰。到 1980 年代中期，那些灰色调色彩看上去有些暗沉，又变化为女性化色彩。1920 年可可·香奈儿（Coco Chanel）将男性服装概念引用到女装上，从此，特别女性化到女装男性化的轮回变化就成为时装舞台中的部分内容。

专业词汇

40. dusty *adj.* 浅灰色
41. elegant *adj.* 优雅的
42. refine *vt.* 精致
43. sophisticate *adj.* 成熟的
44. closet *n.* 衣橱

Passage 6

Fashion Drawing

Drawing may be described as an evolutionary① process that is fundamental to communicating ideas. This is also true of fashion drawing, with its distinctive nuances and associations with style. The exciting breadth[1] and diversity② of what constitutes fashion drawing today is testimony to the creative vision of fashion designers and fashion illustrators[2] alike. It reflects the range and scope[3] of media now available, from a simple graphite③ pencil to sophisticated CAD programs.

1. The sketching process

Fashion sketching not only involves the act of drawing an initial idea but also the process of considering and developing the idea across the pages of a sketchbook. It is always best to have an idea of what you want to draw. This may sound obvious, but fashion sketching should be purposeful, not random[4] or too abstract. In many respects a fashion sketch is a problem-solving process, which brings together the visual elements of articulating an idea in its purest form. This can mean recording a sudden idea before it is lost or forgotten, or capturing a moment in time, such as observing a detail on someone's garment.

A fashion sketch should seek to record and make sense of an idea. This is largely achieved with any one or more of three components: establishing the overall silhouette of a garment or outfit; conveying[5] the style lines of a garment such as a princess seam[6] or the positioning of a dart[7]; and representing details on a garment such as a pocket[8] shape, top stitching[9] or embellishment. Some sketches may appear spontaneous④ or similar to mark making but they should all be linked by a common understanding of the human form and an end use.

Graphite or drawing pencils are ideal for shading and creating variations of line quality. While this is a good way to get started, it is also well worth developing the confidence to sketch with a pen. Sketching in pen requires a more linear[10] approach to drawing, which can often enhance the clarity[11] of a design idea, and it is no less spontaneous than using pencil.

通用词汇

① evolutionary *adj.* 进化的　　　③ diversity *n.* 多样性　　　④ graphite *n.* 石墨
② spontaneous *adj.* 自发的

第 6 课

时装画

绘画可以被描述为渐进的过程，具有交流理念的功能。时装画以它独特的细微特征和风格实现其交流功能。今天，各种类型的时装绘画令人激动，证明了时装设计师和时装插画家创造性的构想，也反映了现在有着应有尽有的绘画媒介，从简单的绘图铅笔到多功能的 CAD 软件。

1. 时装草图

时装草图不仅包括将最初理念画出来的行为，还包括在速写本上不同页面画出思考和发展理念的过程。最好你有想画的理念，这听起来显而易见，但是时装草图应该有目的，不要随意或太抽象。从很多方面看，时装草图就是解决问题的过程，以其最纯粹的形式，集中视觉元素，阐明一种想法。可以记录突发的灵感，以免丢失或忘记，或者及时捕捉一瞬间，例如观察某人服装上的细节。

时装草图就是寻求理念，记录下来，使理念富有含义，草图上包含一到三方面内容。它们是：确定一件服装或一套服装的整体轮廓；表达服装的款式线条，如公主线或省的位置；表现服装的细节，例如口袋形状，表面线迹或装饰。一些草图看起来是即兴的或者似乎是草草几笔，但是它们都必须与大家理解的人体和最终用途相联系。

石墨或绘画铅笔是理想的工具，可以画阴影和各种形式的线条。在开始时就用钢笔是一种好方法，很适合建立绘画的自信。钢笔草图采用线描手法，使设计理念更加清晰，即兴程度不比铅笔弱。

专业词汇

1. breadth *n.* 宽度
2. illustrator *n.* 插图画家
3. scope *n.* 范围
4. random *adj.* 任意的
5. convey *vt.* 传递
6. princess seam 公主线
7. dart *n.* 省道
8. pocket *n.* 口袋
9. top stitching 面上切线
10. linear *adj.* 直线的
11. clarity *n.* 清楚

2. Working drawings

In fashion it is quite usual to produce a series of rough sketches or working drawings in order to arrive at a design or collection proposal. This allows the designer to develop variations on an idea, before making a final decision about a design, whilst at the same time forming part of a critical process of elimination and refinement. The process of reviewing and refining a design involves collating ideas in line-up sheets[12]. These represent drawings of outfits (not individual[13] garments), which are visually presented on the human figure as a coherent statement for a collection proposal. Line-up sheets are more practical than inspiration sketches or rough sketches and are generally clearer to understand on the page. Their primary purpose is to assist with visual range planning and the commercial requirements of formulating[14] ready-to-wear clothing ranges. Consequently, they have no real basis in haute couture or bridal wear, which is more about representing the individual.

3. Sketchbooks

Sketchbooks are the repository of a fashion designer's ideas, observations and thoughts. Whilst there is no template[15] for the perfect[16] sketchbook (and they are not solely the preserve of fashion designers), a good fashion sketchbook should enable the designer to progressively record and document a series of ideas and inspirations through related visual and written material accumulated[17] over time.

All sketchbooks evolve in response to changing influences and circumstances. The true value of a sketchbook is in how the designer uses it to pause and reflect on their work in a meaningful way in order to continue to the next stage of the design journey. It can sometimes be difficult to fully comprehend this when starting out; there may be a temptation to fill up the opening pages with lots of secondary images but this will not lead to a personal sketchbook unless it starts to take on the personality of the user, rather like a personal diary or journal. A sketchbook should become as individual as your fingerprint and provide you with a growing resource from which ideas and concepts can be explored[18] and developed without feeling self-conscious. Sketchbooks also enable you to explore and develop your own drawing style; the book will build up over time and its resource value will increase. One of the most useful aspects[19] of a sketchbook is its portable[20] nature, allowing you to carry it around and enter quick thumbnail sketches or observational drawings.

Most fashion student sketchbooks are A4 size. However, there is no fixed rule on this as some students successfully work with A3-size sketchbooks. Sometimes working across a landscape⑤ A3 format can be useful for sketching A4-size fashion figures and developing preliminary line-ups. The smaller A5 pocket-size sketchbooks can be useful for discreetly⑥ carrying around; they also work well as fabric swatch[21] books and for entering additional thumbnail sketches.

4. The fashion figure

The proportions of a fashion figure are often exaggerated[22] and stylized, particularly for women's wear drawings. This can sometimes be slightly confusing to the untrained eye but in fashion terms

2. 工作画

在时装中，通常要画一系列潦草的草图或工作画，目的是得到一种设计或一个系列的提案，便于设计师在做出最终设计决定之前开拓多种理念，与此同时，也是一种取消和修改设计的重要过程。设计的回顾和修改过程需要在图片旁边配上一张纸条，上面写着关于理念的文字。工作绘画要画整体着装状态（而不是单件服装），用人体形象表示，为系列提案配备文字说明。配上小纸条要比灵感草图或大概的草图来得实际，更容易理解页面上的内容。它的主要目的是帮助制定款式品类和根据商业需求制定成衣的品类。而说到高级时装或新娘服装，就不需要这些，因为它们要体现更多的个性特征。

3. 速写本

速写本，是时装设计师理念、观察和思维的储藏室。但是没有完美速写本的模板（没有针对时装设计师携带的速写本），一本好的时装速写本应该是设计师不断累积图形和文字材料，记录一系列理念和灵感的档案库。

速写本受环境影响不断变化。速写本的真正价值在于设计师如何用这种有意义的方式停下来认真思考他们的工作，以便继续下一阶段的设计。刚开始时有时很难完全理解这种方式，有一种诱惑是用很多二手图片资料填满空白页面，如果这样，速写本就没有个性，除非开始时它就具有使用者的个性特征，而不是像个人的日记或者个人的行程计划。速写本应该像人的指纹那样具有个性，给你提供越来越多的资源，使你能够无意识地从中研究和发展理念和概念。速写本也能使你研究和发展你自身的绘画风格，随着时间的建构，速写本的资源价值也在不断地增加。速写本最大好处是携带方便，不管去什么地方你都可以带着它，快速地画一些大拇指大小的草图和观察到的事物。

时装学生的速写本大多数是 A4 大小。但是没有统一的标准，就像有些学生使用 A3 尺寸的速写本很顺手，有时画风景用的 A3 尺寸速写本画 A4 尺寸的时装人像和加注配文时很有用。口袋大小、A5 尺寸的速写本也是非常有用的，能够很体面地到处携带，它也可以当着面料小样本和画一些拇指大小的草图。

4. 时装人体

时装人体经常比例夸张和风格化，特别是画女装。有时可能被误以为是眼光没有经过训

专业词汇

12. sheet *n.* 纸
13. individual *adj.* 个性的
14. formulate *vt.* 构想出
15. template *n.* 模板

16. perfect *adj.* 完美
17. accumulate *vt.* 逐渐增加
18. explore *vt.* 探索
19. aspect *n.* 方面

20. portable *adj.* 便携式
21. swatch *n.* 样本
22. exaggerate *vt.* 夸张

通用词汇

⑤ landscape *n.* 风景
⑥ discreet *adj.* 谨慎的

it represents a statement of an ideal rather than an actual body shape. This ideal is then aligned to a contemporary look that is viewed through the visual lens of fashion.

Since the late 1960s and 1970s exaggerated proportions have generally prevailed[23] and continue to exert an artistic influence over most fashion drawings. Most standing fashion figures are proportioned between nine and ten heads in height. Most of the additional height is gained through the legs[24], with some added to the neck[25] and a little added to the torso[26] above the natural waist[27]. When drawing the fashion figure the look might refer to the prevailing styles of the season, such as the position of the fashion waist, or it may be an exploration of voluminous or contoured[28] clothing styles with reference[29] to influences from a particularly favoured[30] model or celebrity.

There are fundamental differences between the fashion proportions for drawing men and women. Women's fashion proportions are mostly concerned with extending height through the legs and neck, with the resulting drawings taking on a sinuous[⑦] and gently curved appearance. For men the drawing approach is altogether more angular[31].

5. Drawing from life

Drawing from life, which is an excellent way to develop and refine your drawing skills, involves observational drawing of real-life male or female figures. It is important to consider the appropriate art materials and media, such as charcoal[⑧], pen or pencil, as well as paper type[32] and the eventual scale[33] of work.

Drawing is a process that can be improved[34] and enhanced with regular[35] practice and life drawing offers the particular opportunity of developing and improving hand-to-eye coordination. This is essentially about trusting yourself to spend more time looking at the figure in front of you, rather than by glancing[36] at the figure then looking at the emerging drawing itself and drawing from memory.

This is a common mistake among life-drawing students. It is very important to study the figure before you start to draw. Try to make sure that you are in a good viewing position and then analyse the pose. If the figure is standing it is essential to establish which leg is taking most or all of the weight[37]; this will critically determine the stability[38] of the pose in relation to what is called the "balance[39] line". The balance line is an imaginary line that drops from the base of the centre of the neck down to the floor at the position of the foot[40]. It can be drawn on the paper and used as a guide to ensure that the figure remains standing without "tipping over" on the page. As a general rule, the leg that is supporting the weight of the pose, which should always be drawn before the other leg, will curve down to the floor and should join up with the balance line at the outside edge[41] of the foot.

The principle of the balance line applies to all standing fashion poses including those simulating a walking pose. It is also applicable to menswear although men's poses are generally made less dramatic[42] and gestural than for women's fashion drawing.

Studying the pose first also allows time to evaluate[43] distinctions between the "actual" figure and the expression[44] of an "ideal" fashion figure for womenswear or menswear. Proportions in fashion drawing represent an ideal, so it follows that the life figure does not need to be drawn as an exact representation[45].

练，但是在时装领域它表达了一种理想，而不是现实中人体的体型。这种理想与当代形象相一致，是时尚透视镜里的形象。

自从 1960 年代后期和 1970 年代开始，大多数时装绘画逐渐流行夸张的比例，并受艺术的影响。大多数站立的时装人体高度都在 9 个到 10 个头之间的比例。增加的高度大多数分配在腿部，还有一些在颈部，另外一点点增加在自然腰以上的身躯部位。在画时装人体时，样式应该是这个季节流行的样式，例如时尚腰线的位置，或者是受某个知名模特儿或明星的影响，对服装的体积、轮廓线的研究。

男性和女性的人像绘画比例有根本性的差别。女性时装比例通过延长腿部和增加颈部高度，呈现出蜿蜒柔美的画面，而男性时装画强调棱角。

5. 写生

写生是发展和增强你绘画技能的最好方法，包括对真实男性和女性人体的写生。重要的是考虑用适当的艺术材料和媒介，例如墨石、钢笔或铅笔，不同种类的纸和纸的大小。

通过有规律的写生练习，绘画能力得到提高，同时也是培养手眼协调的最佳机会。从根本上要相信自己，因此花费较多时间凝视你面前的人体，而不是匆匆一瞥人体就画，或凭记忆画。

学生在人体写生时普遍存在错误。在开始画之前研究人体非常重要。你设法使自己处在一个视角好的位置上，然后分析姿势。如果人体是站立的，那么关键是要确定哪条腿承受大部分或全部重量，它对决定姿态的稳定性很关键，与平衡线相关。平衡线是一条想象的线条，从脖颈底部中心，到地板上脚的位置。它可以画在纸上，作为一条引导线，确保人体保持站立，不至于倾倒翻出纸面。一般规则是，姿势中支撑重量的那条腿，要比另一条腿先画，从上到下呈现弧线，平衡线落在脚的周围。

平衡线原则适用于所有站立姿势，包括那些模拟行走的姿势。男性时装画也要考虑平衡线，尽管时装画中男性没有女性的动态姿势夸张。

无论是画女装还是男装，首先要花一些时间研究姿势，评估真实人体和理想的时装人体之间的差别。在时装绘画中比例代表了一种理想，因此不需要精确表现真实形象，它需要可

专业词汇

23. prevail *vi.* 盛行
24. leg *n.* 腿部
25. neck *n.* 颈部
26. torso *n.* 躯干
27. waist *n.* 腰部
28. contour *vt.* 画轮廓
29. reference *n.* 参照
30. favour *vt.* 喜爱

31. angular *adj.* 角度
32. type *n.* 类型
33. scale *n.* 规模
34. improve *vt.* 改进
35. regular *adj.* 有规律的
36. glance *vi.* 一瞥
37. weight *n.* 体重
38. stability *n.* 稳定

39. balance *n.* 平衡
40. foot *n.* 脚
41. outside edge 边缘线
42. dramatic *adj.* 戏剧性
43. evaluate *vt.* 评估
44. expression *n.* 表达

通用词汇

⑦ sinuous *adj.* 蜿蜒的

⑧ charcoal *n.* 木炭

This requires interpretative[46] visualisation, which is an essential release[47] for fashion drawing.

6. Creating poses

Fashion drawings are frequently[48] characterised by gesture[49] and movement[50], both of which are ideally suited to exploration through drawing the fashion figure from life. Part of a fashion drawing's allure is its seemingly effortless style, which is sometimes the result of a careful selection of lines and what is left to the imagination of the viewer. In this regard it is important to note the value of line quality in the fashion-drawing process.

Line quality describes the varieties of drawn lines or marks that have their own inherent characteristics depending on the media that is used, the paper quality, the speed at which the line is made and even the angle of the pen or pencil as it moves along the surface[51] of the paper. Distinct from adding tone and shading techniques, the use of line to convey essential information is integral[52] to most fashion drawings.

Some of the most expressive and visually engaging fashion poses are the result of linear drawings, where selective line quality is used to maximum[53] effect. An understanding of fashion proportions and the standing balance line is essential as a building block[54] for more gestural poses, which instil movement and personality into a fashion drawing. In addition to studying poses from life, it is also possible to develop poses by tracing over figurative photographs in magazines, but this needs to be approached with care: consider the image only as a starting point. Fashion is, after all, a human activity[55] so it follows that developing and creating studied poses is a useful exercise and will aid the development of templates or sketch for future use.

It should be possible to use a pose more than once and for a template figure to present different design ideas. While the pose should be relevant[56] to the context[57] of the clothing (for example, it would make little sense to draw a sporty pose for a wedding dress or an evening gown), creating the pose is much more about the body underneath. Look for movement lines that run through the body—not the outline[58] of the figure—noting the intersections[59] of the pose at the bust[60], waist and hip[61] positions. The leg supporting the weight must be grounded, but the other limbs[62] can be modified or adapted to enhance gestural qualities. In this way, the resulting fashion poses can exaggerate the "actual" to project a more expressive "ideal".

7. Heads, faces and hair

Fashion heads, facial features and hairstyles are worthy of special consideration in fashion drawing; they can convey a multitude of essential style and gender[63] information.

The very personal and unique attributes[64] that a face can contribute to a drawing are worth exploring through practice and exercises. Much like the evolution of fashion drawing itself, the "ideal" face changes over time and takes on many guises. Make-up trends continue to have a direct influence on contemporary fashion faces and it is always useful to collect magazine tear sheets from which to study and evaluate different faces and proportions.

解释性的视觉效果，这就是时装绘画表现的关键所在。

6. 创造姿势

时装画通常以姿势和动态为特征，两者最适合从时装画的写生中探索。一些时装画的魅力是看上去很轻松，实际上线条经过精心计划，留给观者去想象。从这方面来看，在时装绘画过程中线的特征有很重要的价值。

线的特征就是绘画中线的变化，或者是使用不同媒介体现出的内在特征。例如纸张的质量、画线的速度，甚至钢笔或铅笔在纸表面上移动时的角度等。与增添调子和阴影技巧不同的是，线条表达了本质信息，大多数时装绘画都少不了线条。

一些最具表现力和最具视觉吸引力的时尚姿势，就是线描的结果。通过选择不同特征的线条，取得了最佳的效果。为了画出更多的动态，了解时尚比例和站姿平衡线是很重要的，并且将它们渗透到时装画的动感和个性特征中。此外，从生活中和杂志上的人像照片都能研究姿态，但要谨慎对待：研究图像只是出发点。毕竟，时尚是人的一种活动，应该不断地研究和创新姿态，才是有效的锻炼，有利于发展时装画模板或速写，为将来所用。

一种姿势可以不止一次使用，一种人像模板，可以表现不同的设计理念。但是姿势应该与服装的内容一致（例如，采用一种运动性姿势表现婚纱或晚装是毫无意义的），创新姿势时要更多地考虑服装下面的身体。寻找贯穿身体的活动线条——不是身体的轮廓线，注意胸部、腰部和臀部位置横断面的姿势。支撑身体重量的腿必须着地，另外一条腿用来修饰，增强姿势的美感。用这种方式，可以夸张时尚姿势，将实际形象表现为更加理想的形象。

7. 头、脸和头发

在时装画中，时尚的头部、脸部特征和发型值得特别关注。它们可以表达多种基本风格和性别信息。

通过实践和练习，探索脸部的个性化和独特特征是非常值得的。就像时装画自身的变化那样，理想脸形也是随着时间而变化的，并采用不同的化妆。化妆潮流不断地对当代时尚的脸型产生直接影响。从杂志上撕下的脸型小纸片汇集在一起，便于研究和评估各种脸型和比例。

专业词汇

45. representation *n.* 表现
46. interpretative *adj.* 解释
47. release *n.* 释放
48. frequent *n.* 频繁的
49. gesture *n.* 姿势
50. movement *n.* 运动
51. surface *n.* 表面
52. integral *adj.* 整体
53. maximum *adj.* 最大值
54. block *n.* 块面
55. activity *n.* 活动
56. relevant *adj.* 相关
57. context *n.* 上下文
58. outline *n.* 轮廓线
59. intersection *n.* 交叉点
60. bust *n.* 胸
61. hip *n.* 臀
62. limb *n.* 肢体
63. gender *n.* 性别
64. attribute *n.* 特征

Although faces can be drawn in the linear style that is so often used in fashion, they can also lend themselves to applications of tone and shade. Structurally, the forward-facing head is oval in shape for women—much like an egg shape—and should be horizontally[65] intersected at mid-point[66] to position the eyes. The mouth[67] is usually arranged halfway[68] between the eyes and the base of the chin. The mouth could be considered in two parts with its upper and lower lips[69]. The upper lip should include an 'M' shape definition[70]. The nose[71] may either be represented with dots for the nostrils[72] above the top lip of the mouth or with an added off-centre[73] vertical[74] line from the front of the face as if to indicate a shadow[75]. Noses are rarely given any prominence[76] in fashion faces as the eyes and lips become the main features.

The ears may be discreetly added at the side of the head starting at eye level[77] and ending just above the nostrils; they can be useful for displaying earrings[78], if appropriate.

The hair should be carefully considered as this can have a transforming effect on the appearance of the fashion head. Again, collect tear sheets from magazines in order to build up a visual file of hairstyles as it can be quite challenging to imagine them without a reference point and of course, hairstyles for women vary enormously. If it is visible the hairline[79] should be drawn around a quarter of the way down from the top of the oval shape of the head, line, shade and colour can all be added according to the style requirements.

8. Arms, hands, legs and feet

When drawing a fashion figure it is important to consider the hands, arms, legs and feet in relation to the pose and gestural qualities.

When drawing an arm, consider it in three parts: the upper arm[80], the elbow[81] and the lower arm[82]. The upper arm is attached to the shoulder[83] from which it may pivot[84] depending on the angle of the torso. It has a smooth, gently tapering[85] upper section[86] that reaches down to the elbow position. The lower arm tapers more visibly to where it joins the hand.

The hands have two main parts: the front or back of the palm[87] and the fingers[88] and thumb[89]. Both parts may be elongated[90] to offer the fashion figure a range of gestures and actions, which will all enhance the drawing. Consider the angle of the lower arm when drawing the hand. Fingernails may be included but knuckles are not usually emphasised[91]: too much detail on a hand can make it look wrinkled[92]. You could also try drawing the hand resting on the hip with the fingers hidden from view.

The feet are usually drawn in a simplified way that mostly assumes a shoe[93] line. When starting out it is helpful to practise sketching bare feet, but the foot will usually be hidden from view within a shoe, which can be drawn in a huge variety of styles. The overall look will be determined by the angle of the foot and whether or not the shoe has a heel.

As fashion drawing is largely concerned with presenting an interpretation of an ideal figure rather than realistic proportions, so it follows that drawing the legs is an exercise in artistic licence. Fashion legs are routinely extended in the upper leg and thigh, and below the knee[94] to where the ankle[95] meets the foot.

时装画中经常用线画脸部，还可以用色调和阴影来润饰它。从结构上来说，女性正面头型是椭圆形，很像鸡蛋的形状，眼睛在水平中点位置上。嘴在眼睛和下颌底部的中间位置。嘴由上嘴唇和下嘴唇两部分构成。上嘴唇呈"M"形状。鼻子在上嘴唇上方，用点表示鼻孔，或者在脸部中间增加一条偏离中心的垂直线来表示阴影。时装画中，很少突出鼻子，而是突出眼睛和嘴唇。

在头的两侧谨慎地添加耳朵，上端与眼睛平齐，下端对齐鼻孔上方。如果需要的话，画上耳环是很有用的。

头发应该仔细考虑，因为它能够改变时尚头型外貌。同样，从杂志上收集发型图片，可以建立发型的视觉档案，因为没有参考资料的想象很艰辛，当然，女性发型千变万化。如果发际线位置明显的从椭圆形头顶向下四分之一处，那么根据风格添加线条、阴影和色彩。

8. 手臂、手、腿和脚

在画时装人体时，重要的是还要考虑手、手臂、腿和脚与姿态特征的关系。

在画手臂时，将手臂分成三个部分：上臂、肘部和前臂。上臂与肩部相连，根据躯干的角度可以旋转。从上部分向下到肘部逐渐变细，前臂到手的部位变得更细。

手由两部分构成：掌心、手背和四指、拇指。这两个部位都可以拉长，为时装人体提供各种姿势和动作，为此增加绘画的效果。画手时，要考虑前臂的角度。可以画指甲，但不要强调指关节，手上太多的细节，使手看起来布满了皱纹。也可以尝试着将手画成搭在臀部上，手指隐藏起来看不见。

脚经常使用简单的方式来画，主要画鞋子的线条。开始学画时装画时，画光脚是非常有用的。但是，脚通常藏在鞋子里，鞋子可以画成很多种样式。脚的角度决定鞋子的整体形状，决定是否画鞋子的后跟。

因为时装画主要关注理想形象的表现形式，而不是真实的比例，因此画腿部时要从艺术的角度来画。按常规，时尚的腿在腿上部和大腿处、膝盖下方到脚都要延长。

专业词汇

65. horizontally *adv.* 水平地
66. mid-point *n.* 中点
67. mouth *n.* 嘴
68. halfway *adv.* 中间
69. lip *n.* 嘴唇
70. definition *n.* 定义
71. nose *n.* 鼻子
72. nostril *n.* 鼻孔
73. off-centre 偏离中心
74. vertical *adj.* 垂直
75. shadow *n.* 阴影
76. prominence *n.* 突出
77. level *n.* 水平
78. earring *n.* 耳环
79. hairline *n.* 发际线
80. upper arm 上臂
81. elbow *n.* 肘部
82. lower arm 前臂
83. shoulder *n.* 肩部
84. pivot *vi.* 枢轴
85. taper *vt.* 逐渐变细
86. section *n.* 部分
87. palm *n.* 手掌
88. finger *n.* 手指
89. thumb *n.* 拇指
90. elongate *vt.* 延长
91. emphasise *vt.* 强调
92. wrinkle *vi.* 起皱纹
93. shoe *n.* 鞋
94. knee *n.* 膝盖
95. ankle *n.* 脚踝

Passage 7

Fashion Show

A catwalk fashion show is a sales promotion[1] mechanism in the clothing industry and a widely recognized cultural event. Although the fashion show is essential to how the fashion industry works, it has also become a cultural icon in its own right.

A fashion show is a biannual presentation of a new clothing collection on moving bodies for an audience. A new collection is produced by a designer, brand company, or group of companies. The parade of moving bodies makes up an essential feature of a fashion show, and has given rise to the modelling profession as well as to a range of conventions of movements, poses and looks. It is accompanied by music which emphasizes the rhythm of movement and blocks out other sounds from the overall impression. The moving bodies are predominantly female. Although menswear fashion shows have been held since the late 1920s, they are still by far outnumbered by women's wear shows.

1. Framing the fashion show

A fashion show and its modes of presentation may be explained to a large extent in terms of frame analysis. This applies both to the spatial (setting, catwalk, set and runway design), and temporal (music, performance, staged appearances) framing of the fashion show (see Figure 7.1). Framing devices include the technologies, props and conventions that set the fashion show apart from ordinary interaction and define what is going on both within the fashion show itself and between the fashion show and the outside world.

Firstly, fashion shows are set apart from the outside world in terms of their location. As part of a Fashion Week programme, fashion shows are often held in conjunction with trade fairs in exhibition grounds that are typically (but not always) located on the outskirts of large and medium-sized cities. The atmosphere in such locations (whether they be exhibition hall or marquee tent) is neutral and anonymous[2]. Typically, they have no windows and the fact that they are totally enclosed enables the staging of the fashion show to be completely controlled. In this way, the attention of the invited

---

专业词汇

1. promotion *n.* 促销   2. anonymous *adj.* 匿名

第 7 课

时装秀

T台时装秀是服装行业的促销手段,也被广泛认为是一种文化活动。尽管时装秀本质上是展示时装企业的工作,但它自身已经具有文化的象征含义。

时装秀就是通过模特儿的移动,为观众展示一年两次的服装新品。新的服装产品由设计师、品牌公司或公司集团生产。活动的模特儿展示,构成了时装秀的重要特征,因此产生了模特儿职业和一系列约定俗成的走步方式、姿势和外貌。为了强调运动的节奏感和全神贯注免受其他声音的干扰,需要配有音乐。模特儿主要是女性。尽管从1920年代后期,已经举行男性时装展示,但它仍然不及女性服装的时装秀多。

1. 时装秀的组织结构

时装秀和它的展示模式很大程度上可以用结构分析来解释。时装秀采用了空间性(背景、T台、座位和T台设计)和瞬时性(音乐、表演、舞台形象)这两种结构(见图7.1)。结构设置包括技术、道具和一些常规做法,使时装秀与正常的互动活动不同,定义了正在举行的时装秀本身,也将时装秀与外部世界隔离开来。

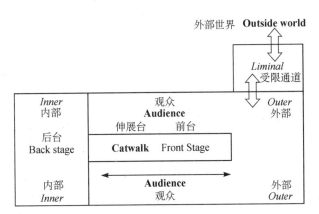

Figure 7.1　The fashion show framework
图 7.1　时装秀场框架

首先,举办时装秀的地点与外界隔开。作为时装周的部分日程,时装秀通常在展览馆的贸易展销会上举行,一般(但不总是)在大型和中等城市的郊区。在这些地方(展厅或大帐篷)的气氛一般没有倾向性和特色。通常没有窗户,事实上它们是完全封闭的,时装秀的表

audience is directed away from the outside world and made to focus entirely on the ephemeral[3] setting that frames the fashion show performance.

In addition to this type of neutral setting, fashion shows are also held in locations that are chosen to colour the atmosphere of the show. In French and Italian fashion shows aristocratic ancient régime palais may be selected, while other typical locations include derelict factories, warehouses[4], theatres and museums. In reality, therefore, fashion shows are held in all sorts of locations.

In the golden age of haute couture, the stage was referred to as the podium. Since then, other terms have taken over, including catwalk, signifying a narrow passage, and runway, which — with a reference to the takeoff of an airplane — refers to the launching of a new collection. The raised dais — like a theatre stage, college high table or church altar — gives ritual significance[5] to the activities performed, and exalts the persons performing, there, thus separating the audience from the performers, those who look from those who are looked at. The direction of gazes is reenforced by lighting that bathes the runway in strong light and leaves the surrounding audience in the dark.

Not all fashion shows make use of a raised stage. Instead, they create a catwalk by making an aisle between audience seats or in other ways use the features of the location to create a space visibly laid out for the parade. Invariably, the fashion show makes associations to other situations where people walk along aisles between seats. In the heyday[6] of Paris haute couture, at the end of the parade the male creator would accompany a model wearing the bridal dress, traditionally the last number in a fashion show, in a gesture that quoted the convention of the father leading a young woman up the church aisle at her wedding.

When it comes to the placement of the audience around the stage, we find a whole set of framing conventions reflecting what Dorinne Kondo has referred to as the politics of seating. Invariably, photographers are placed at the end of the runway to enable long-lens shots[7] of the models walking down the catwalk. Depending on the importance of the show, the crowd of photographers may vary from a few to a veritable forest of telephoto lenses and cameramen, although for TV, webcast transmissions or videotaping, two to three cameras give the best coverage. This means that space needs to be available for not only a head-on spot, but also a side view and a position closer to the start of the runway for the "return" shot. Photographers also need to have access to positions from which to shoot the guests, especially the "dignitaries" in the front-row[8] — if not during the show itself, then immediately before or after it. So photographers are given privileged visual positions, underlining the importance of the mediation of the event to audiences not present.

In the politics of seating, choice spots are determined by their proximity① to, and view of, the action on the catwalk. The seating area in front of the cameras at the end of the stage is considered to provide the best view when available. In general, though, the best seats at a fashion show are in the front row at the end of or along the stage. These front row seats are reserved for the most important guests, such as magazine editors, who are the essential filters through which the shows are reported in the media, and celebrities, whose presence may add prestige to the show. In sales shows, buyers are also seated in the front row, but today most buyers will view the collection informally in the

演完全在控制之中。用这种方法，受邀观众的注意力远离外部世界，完全聚焦于时装秀转瞬即逝的表演情景中。

除了在没有倾向性的地点之外，时装秀也选择能为秀渲染气氛的地点举行。在法国和意大利也许会选择在具有贵族气息的古老宫殿里举行时装秀，其他一些典型的地点包括废弃的工厂、仓库、剧院和博物馆。事实上时装秀可以在各种地点举行。

在高级时装的黄金年代，舞台称为表演台。从那以后，其他一些术语取代了这个名词，包括猫步舞台，即一条窄的通道和跑道，引自飞机起飞的跑道，意味着新系列服装的推出。升起的舞台——像剧院的舞台、学院高的讲台或者教堂的圣坛——使表演活动富有仪式的含义。表演者高度的提升，与观众区分开来，表演者与观众彼此被对方观看。T台上沐浴着强烈的光线，光照加强了凝视的方向，使T台周围的观众处在黑暗中。

并不是所有的时装秀都采用高出地面的舞台，而是把观众座位排在两边，中间形成一个甬道成为T台，其他方法就是使用地点特征，创造一个可以走步的视觉空间。不变的是，时装秀总是在两排座位之间设置可以走步的甬道。在巴黎高级时装鼎盛时期，时装秀最后压台的是一位男模陪伴一位身穿新娘婚纱的女模走到T台的末端，引用了新娘父亲搀着新娘在教堂甬道上行走的传统形式。

当谈到舞台周围观众的位置时，我们发现了一整套惯常的组织结构，Dorinne Kondo 称之为政治化座位。毋庸置疑，摄影师被安排在T台的末端，使他能够用长焦镜头拍摄正在T台行走的模特。摄影师的人数由时装秀的重要程度而定，或许只有几个，或许像森林般的长焦镜头和摄影人。虽然是为电视、广播的传输或录像拍摄，只有两到三台相机有好的视角，这意味着要留有正前方、两侧的空间供拍摄，以及T台开始的位置，拍摄模特返回的镜头。摄影师还需要有一个能拍到贵宾的位置，特别是指定坐在第一排的那些名流。如果不在秀场里，也要在秀场开始前或秀场结束后赶紧拍摄他们。因此，摄影师被给予视角位置特权，强调了摄影师对秀场的不在场观众所起的主要宣传作用。

在政治化座位中，座位安排由与T台上活动的远近程度和视野决定。在正对T台、照相机所处位置的前排座位有最好视野。一般来说，时装秀中最好的座位是T台末端和T台两侧的第一排座位。这些前排座位预留给最重要的贵宾，例如杂志编辑，媒体报道的秀场信息要经过他们的筛选；明星，他们的出席为秀赢得了声誉。在销售性的时装秀中，买手也坐在前排，但是，今天大多数买手都是在非正式的展示厅里观看新品系列。因此，越来越多的时装秀是对媒体展示完整的形象，只是间接地为买手服务。前排后面的位置安排不太重要的宾客，

专业词汇

3. ephemeral *adj.* 短暂的
4. warehouse *n.* 仓库
5. significance *n.* 意义
6. heyday *n.* 盛世
7. shot *vt.* 拍摄
8. front-row 前排

通用词汇

① proximity *n.* 接近

showroom, so that the purpose of the fashion show is increasingly to present an overall[9] image for the press[10], and only indirectly for the buyers. The seats behind the first row are for less important guests, including many buyers and business contacts, company employees, design school students, and other members of the public interested in attending the fashion show. In large fashion shows there may be a standing area behind the VIP seats. In other shows, the first row is extended by manipulating the space, so that everyone in the audience can have a first row seat. This is possible in fashion shows held in large premises where the catwalk area can be extended, sometimes through several rooms and corridors. It can also be done if the parade of models trails around or through audience seating arrangements.

At the back of the catwalk is the set design, which serves as the backdrop of the performance. A fashion show is typically accompanied by a set of slides, projecting the logo[11] and credits[12], as well as images, colours and designs that enhance the concept of the show. The set design also separates front stage, where collections are appreciated and consumed by the audience, from back stage, where they are pieced together and produced by the designer concerned. As such it marks the point where models change their staged pace as they prepare to leave or enter the front stage theatre. While the front stage is carefully scripted in its staged framing devices, both in place and time, in order to exclude all possibility of unscripted behavior and individual improvisation in the ritual performed, the back stage consists of ordered chaos—order in the necessary arrangement of clothes enabling models to hurriedly dress, change and dress again, but chaos in the sheer number of different kinds of personnel present and the multiplicity[②] of tasks that they must carry out to enable the front stage performance to take place.

In this framework, the fashion show can be said to consist of two performances encased in each other. The first one starts with the arrival of the audience, which is obliged to form a queue to enter a single access point to the fashion show stage (often via a liminal space between the outside world and the show venue), and every member of which is vetted and passed or rejected by gatekeepers who examine printed invitations and check individual names as printed on their invitation lists. The start of the show is almost invariably delayed, which incidentally gives everyone time to observe the crowd and spot which editors and celebs grace the show with their presence. VIP guests may calculate the delay and time their arrival at the venue accordingly, with the more famous being allowed to arrive later than the hoi polloi.

The second performance, the performance of the models on stage, starts with the outbreak of music — usually so loud that it drowns all other sounds — together with an adjustment of lighting. It is at this point that the first model appears on stage. The music accompanying a fashion show is selected and played by a DJ in order to match the designer's concept for the show. Together music, lighting and slides are used to emphasize discrete sections in the collection presented. The fashion show usually lasts for no longer than fifteen to twenty minutes. Its end is signified by the appearance of all the models who parade together down the runway to the accompaniment of the audience's applause. Eventually, the designer whose collection has been shown also makes an appearance, sometimes brief and informal, sometimes obviously choreographed[③]. Not infrequently, a few

包括很多买手、生意往来者、公司员工、设计学校的学生和其他一些对时装秀感兴趣的公众。大型时装秀，在VIP座位后面还有站位。另一些时装秀，通过空间设置延伸前排座位，使每位观众都在前排座位。如果在一些大型场所举办时装秀，有可能将T台区域延伸，有时穿过几个房间和走廊。同样，如果模特沿着或穿过观众的座位行走，效果也是一样。

T台后面是背景设计，衬托表演。时装秀通常配一组幻灯片，包括商标和企业的荣誉、图像、色彩和设计，强调时装秀的概念。背景设计将舞台分为前台和后台。前台是为了观众欣赏和购买发布会的产品；后台是根据设计师的意图将单件服装组合并放置在一起。背景也是模特准备离开后台或进入前台的标志点，在这点上，模特走路的步伐开始改变为舞台步伐。为了排除正式表演中所有未计划的个人即兴行为，前台台面从步伐到时间都经过精心设计。后台有序而混乱，服装必须有序摆放，才能保证模特着急匆匆地穿、脱和更换服装。混乱是因为后台有各类人员，他们各自的任务不同，确保前台的表演顺利进行。

在这种框架结构中，可以说时装秀由两个互相包含的表演部分构成。第一部分是由观众的到达开始，他们被要求排队从单一的通道进入会场（通常很小的空间连接外部与秀场），每一个人都要被检查，门卫检查印刷的请柬，核对姓名是否在他们邀请的名单上，决定放行还是拒绝。时装秀开始的时间总要被推迟，正好留出时间每个人可以观看其他人，目睹编辑和名流们的出席给秀的增色。VIP贵宾可能计算延迟的时间和他们到达的时间，越是出名的人越是可以比普通人迟到达。

第二部分表演是舞台上模特儿的表演，随着音乐突然响起——声音之大淹没了其他声音——同时调整灯光。正在此时，第一个模特儿出现在舞台上。时装秀的配乐由DJ师挑选和演奏，要与设计师为时装秀设定的概念相吻合。音乐、灯光和幻灯片加在一起，使发布会呈现的离散部分得到强调。发布会持续的时间是15～20分钟。所有模特儿一齐出现在T台上，表示秀结束，并伴随观众的掌声。最后发布会的设计师亮相，有些人简短而不正式，有些人明显设计了舞蹈动作。通常有些观众到T台前给设计师送花。在这之后，时装

专业词汇

9. overall *adj.* 全部的
10. press *n.* 新闻报道
11. logo *n.* 标志
12. credit *n.* 信誉

通用词汇

② multiplicity *n.* 多样性
③ choreograph *v.* 设计舞蹈动作

members of the audience will come up to the catwalk to hand a bouquet of flowers to the designer. After this, the fashion show has ended and the audience leaves. For many, fashion shows are part of a busy fashion week schedule[13], so they may well be rushing on to the next appointment.

2. Backstage production

Backstage, a large number of people work to realize the show. A relatively basic fashion show involves around twenty people—excluding models and support personnel such as caterers and drivers—and can easily run to a budget of € 60,000. By comparison, for top designer shows, such as those by Dior or Chanel, figures of five million dollars are quoted. In spite of the variations, which do occur, there are bundles of tasks and lines of command that are common. They make for a routinized[④] interaction which is necessary for the success of an event that is usually produced under considerable time pressure.

In fact, preparations start well in advance of the fashion show. Typically, a designer or the fashion house concerned approaches an event agency[14] six months before the planned show to talk about concepts and budgets. The event maker or art director of the event agency presents a concept, which is perhaps modified, but otherwise accepted by the company, and the event maker will then start the actual preparations for the show. In the proposal, some of those who will be involved in the production are named—for example, the stylist, an interior decorator, possibly a photographer to document the event, the production manager, persons in charge of lights and sound, and possibly one or two top models. These people will have been approached in advance and asked to join the project[15]. Upon acceptance, the professionals discuss and come to agreement on the proposal, often supplementing details in their own area of expertise[16]. The involved parties will then prepare their own part in the production, and the art director will present additions or changes to the fashion house for approval. Next a venue is chosen. Normally, it has to be coherent with the theme of the show, unless the latter takes place in a venue set up to house different shows—for example a tent connected with a fashion fair[17]. If this is the case, the following account will need to be modified since lighting, sound, decorations and so on will for the most part already have been put up.

The backstage venue consists of separate stations where the different professionals have their base. This is in order to organize the somewhat chaotic ad hoc workspace, and to make communication easier. Hair and make-up are located in a faraway corner; stylist at the entrance to the runway; people in charge of sound, lighting and decoration all around the front area (see Figure7.2). Most professionals have assistants, and throughout the day, different people arrive and begin their separate jobs. As it is not necessary for the models or dressers to be there when lighting and chairs are being set up, models, for instance, arrive only when the hairdresser, make-up artist, and stylist are ready to start preparing them. Similarly, waiters and seaters will not be present until shortly before the invitees arrive. Depending on the type of fashion house and the scale of its show, the designer or designers will not be at the venue until shortly before the guests.

秀结束，观众离开。对很多人来说，时装秀是时装周繁忙日程的一部分，因此，他们也许会匆忙地赶到下一个预约地点。

2. 后台工作

后台有很多人为秀的举行而工作。相对简单的秀大概需要20人——不包括模特儿和一些协助人员，例如饮食提供者和驾驶员——经费预算大约需要6万欧元。与高级设计师的秀相比，例如香奈儿或迪奥，需要500万美元预算。尽管这种差异存在，但是任务的多少和逐层控制程序是相同的。要制定合作路线图，在时间紧迫的压力下，这是活动取得成功必须要做的。

事实上，在举行时装秀很久之前准备工作就已经开始了。通常设计师或时装屋在举行秀场的前6个月就开始与代理接洽，讨论方案和预算。代理公司的制作人和艺术指导提出一种方案，公司或许会提出修改意见，或许会直接同意，然后制作人就开始为秀进行实际的准备工作。在这种方案下，要邀请一些人参与，例如，造型师、室内装饰师，可能的话要请一名摄影师为此项活动做一些档案的拍摄工作，产品经理、灯光和音响负责人、或许还有一两名超级模特。事先与这些人商量，邀请他们参加这个项目。接受邀请之后，他们开始专业性的讨论，对提案取得一致意见，通常各自从自身专业角度提出细节方面的补充意见。然后，这个团队开始准备自身的那部分工作。若有补充或变化，艺术指导要得到时装屋的认可。接下来就是选择场地。一般来说，地点要与发布会的主题相一致，除非是在一个举行多场秀的地点——例如时装贸易会设立的帐篷。在这种情况下，要做的事情就是修改已经安装好了的灯光、音响、装饰等等。

后台分为几个不同的专业性区域，使得有点混乱的临时工作场所条理化，以便彼此的沟通容易些。发型师和化妆师在后面角落里；造型师在T台的入口处；负责音响、灯光和装饰的人在前面区域（见图7.2）。大多数专业人士都有助手，不同的人在一天中不同的时间到达，准备他们各自的工作。例如，安装灯光和椅子的时候不需要模特和穿衣工。模特儿是在发型师、化妆师和造型师开始为他们做准备时才到达。类似地，服务者和座位引导员只是比客人早到一会儿。根据时装屋的类型和时装秀的规模，设计师或设计师们比客人早到一会儿。

专业词汇

13. schedule *n.* 时刻表
14. agency *n.* 代理
15. project *n.* 项目
16. expertise *n.* 专家
17. fair *n.* 商品交易会

通用词汇

④ routinize *vt.* 使惯例化

Throughout the day, rehearsals[18] are carried out. This is mostly to estimate the time, and make sure that models, choreographer, lighting and sound technicians know when and what to do. Simultaneously, clothing might be stitched up or ironed[19], lights may be put up, and seating arranged. Dressers receive their instructions and, as each dresser often dresses up more than one model, and each change must be done in a few seconds, all the clothes are hung up unzipped and unbuttoned in exact order. The models are introduced to their dressers, and during the event, they will go from the stage to the dressers and wait for them to finish dressing the previous model. To help organize outfits, a photograph of each model is put up at each wardrobe station. At big shows there is a table of order of outfits that serves as a visual reference for all details.

By the time the show is ready to begin, the backstage area will be filled with people who are all connected to the realization of the event. As the venue is often small, only people who are essential are allowed into the backstage area. The production manager is now very important since the level of concentration in each area makes him or her the only link between the different professionals, and the only one who has an overview of the entire production. Hair and make-up are now finished, and are on standby for touch ups during the show. The people in charge of music and lighting are present in the front stage area to keep an eye on the show while it is in progress. As the guests enter the venue, the boundary between backstage and front stage is strictly upheld.

During the show, the production manager acts as the link between front and back stage, cueing lights, sound and the choreographer. The models are lined up just behind the curtain[20] waiting, dressed in the first outfit they are to present. They are cued by the choreographer and enter the catwalk, and as they return, the dressers are waiting with the next outfit ready and unzipped. The models are quickly dressed and sent to the stylist and make up for touching up before going out on the catwalk again. Everything is quickly dismantled after a show. Hair and make-up pack their things; models rush off to another show; and stylists and dressers organize the clothes on hangers[21] or in boxes to be sent to a showroom.

专业词汇

18. rehearsal *n.* 排练
19. iron *vt.* 熨烫
20. curtain *n.* 窗帘
21. hanger *n.* 衣架

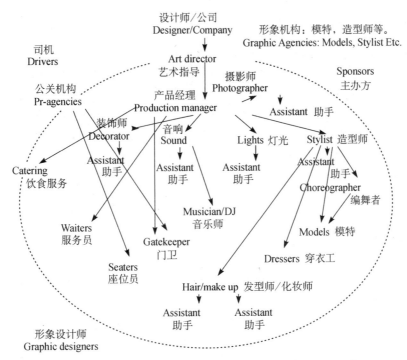

Figure 7.2　Fashion show organisation
图 7.2　时装秀场组织分布

　　在白天要进行排练以计算准确的时间，使模特儿、编舞者、灯光和音响技术员知道什么时候该做什么。同时，服装需要缝合或熨烫，灯光要测试，座位要排放，穿衣工要接受指导，因为每一个穿衣工要为多个模特儿穿衣，每套服装在几秒钟之内更换完成。所有服装敞开拉链和扣子，按次序准确地悬挂在衣架上。模特儿与他们的穿衣工见面，在发布会的时候，他们从 T 台走向他们的穿衣工，等候前面模特儿穿好衣服后为他们更换服装。为了便于服装的搭配，每个模特儿的衣服站点都张贴了此模特事先排好的一组照片。大规模的秀场，有一整套服装搭配的次序表，可以根据图片参考所有细节。

　　时装秀即将开始时，后台挤满了为秀服务的人。因为后台很小，只有那些关键人物允许进入后台。现在产品经理相当重要，因为每个工作站点集中于自己的工作，他或她成了唯一连接各个站点的人，也是唯一能够看到整个秀场的人。发型师和化妆师已经完成工作，站在一旁等待演出时的补妆。负责音乐和灯光的人在舞台前面，不断注视秀的进展。当贵宾进入秀场后，后台和前台被严格控制。

　　在时装秀进行的时候，产品经理起着联结前台和后台的作用，要给灯光、音响和编舞者发出信号。模特身穿要展示的第一套服装排在幕布帘的后面等候出场。他们在编舞者的暗示后，进入 T 台，当他们返回时，穿衣工已经等待在那里，为他们脱掉服装，换上准备好的下一套服装。模特儿被很快地换好服装、送到造型师那里，再次进入 T 台前要补妆。秀结束后，所有东西很快就被撤走。发型师和化妆师打包他们的东西；模特儿匆忙离开去另一个发布会；造型师和穿衣工将服装用衣架整理好，或放进箱子里，然后服装被运送到展示厅。

3. The presentation of a new collection

A collection is a series of garments that a company presents to the market all at the same time. There are a number of fashion shows that do not present a single brand or designer, but a group thereof. Group shows include graduation shows, staged by most design schools where each graduate will present a small collection, typically consisting of three to eight outfits. Similarly, fashion show contests, usually for young designers, consist of multiple small collections, produced under conditions specified by the content organizers. Apart from the entertainment[22], the purpose of such shows is not to sell clothes, but to showcase the capabilities[23] of individual designers, both for the press and for potential employers. In addition, fashion fairs often open with a trend show, which presents a selection of garments by the exhibitors at the fair.

4. On living bodies

Whether a show is an haute couture, ready-to-wear, group, or trend show, the clothes are presented on living bodies.

As moving images became an integral part of fashion show presentations, a strange interplay of movement and montage was created. The staccato[⑤] movements of the models suggested a montage of still images, and a certain number of end poses were in play as models rested momentarily at the end of the runway, offering opportunities for the perfect image for photographers. The movements rather resembled those of soldiers on parade; but while the movements of soldiers are choreographed to be firm, determined and very abrupt[24], the catwalk movement for especially female models had to be soft, swaying[25], and spherical, to reflect particular ideals. Thus, the upper body is kept erect[26] and passive[27], with the arms dangling[28] carelessly along the sides while maybe holding on to a bag[29]. The knees are lifted higher than in a normal walk, with each leg swayed exaggeratedly over the other, as the moving foot is placed in front of the one behind. All in all, this movement pattern causes the model to look like an idealised or stylised object[30].

3. 新系列时装发布会

一个系列就是公司在同一时间呈现给市场的一组服装。有很多时装秀不仅仅展示单个品牌或单个设计师品牌，而是很多人的作品。这样的秀包括很多设计学院举行的毕业秀，即每个毕业生展现数量少的系列作品，通常三到八套。类似地，为了竞赛举行的秀有很多年轻设计师参加，风格各异，每个系列数量较少，是根据主办方的要求进行的设计。这种秀的目的不是为了销售服装，而是为了娱乐、对媒体或想招聘新人的企业展示设计师的个性能力。此外，时尚交易会经常举行趋势发布会，其服装由交易会承办方在交易会的摊位上选取。

4. 活动的人体

不管是高级时装秀、成衣秀、群组秀、还是趋势秀，服装都采用活动的人体展示。由于活动的形象是呈现时装秀的关键，一些奇怪的相互作用的动作和蒙太奇形式被设计出来。模特的一些不连贯动作，就像一系列静态图形的蒙太奇，在T台的末端模特摆放很多姿势，以便摄影师有机会拍摄完美的形象。这种运动很像士兵被检阅时的步伐，但是士兵的步伐被设计成坚定的、坚决的和相当突兀的。而T台的步伐，特别是女性模特，必须是柔软的、前后摇摆的和球状的，表现一种特定的理想形象。因此，上身保持直立和被动，手臂在两侧无意地摇摆，也许拎着一个包。膝盖比正常走路时抬得要高，一条腿夸张地摇晃着跨到另一条腿前面，移动的脚落在另一只脚的前面。总之，这种运动的图式使模特看上去像一个理想化或风格化的物体。

专业词汇

22. entertainment *n.* 娱乐
23. capability *n.* 才能
24. abrupt *adj.* 唐突的
25. sway *vt.* 摇摆
26. erect *adj.* 直立的
27. passive *adj.* 被动的
28. dangle *vi.* 悬荡
29. bag *n.* 包
30. object *n.* 物体

通用词汇

⑤ staccato *adj.* 不连贯的

2

Part two

第二部分

Passage 8

Pattern cutting

It is important for designers to understand as early as possible how a garment grows from a two-dimensional concept into a three-dimensional object. A pattern is a flat paper or card template, from which the parts of the garment are transferred[1] to fabric, before being cut out and assembled.

A good understanding of body shape and how body measurements transfer to the pattern piece is essential. The pattern cutter[2] must work accurately① in order to ensure that, once constructed, the parts of fabric fit together properly and precisely.

1. The block

A block (also known as a sloper[3]) is a two-dimensional template for a basic garment form (for example, a bodice[4] shape or fitted skirt) that can be modified into a more elaborate② design. Blocks are constructed using measurements taken from a size chart[5] or a live model, and do not show any style lines or seam allowance. Blocks must, however, include basic amounts of allowance for ease[6] and comfort; for instance, a tight-fitting bodice block would not have as much allowance added into the construction as a block for an outerwear garment might. A fitted bodice block would also have darts added into the draft[7] to shape the garment to the waist and bust, whereas a block for a loose-fitting overcoat[8] would not need these.

2. The pattern

A pattern is developed from a design sketch using a block. The designer or pattern cutter will add to the block by introducing style lines, drapes, pleats[9], pockets and other adjustments[10] to create an original pattern.

The final pattern features a series of different shaped pieces of paper that are traced[11] on to fabric and then cut out, before being seamed together to create a three-dimensional garment.

Each pattern piece contains notches[12] or points[13] that correspond to a point on the adjoining pattern piece, enabling whoever is making the garment to join the seams together accurately. The pieces need to fit together precisely, otherwise the garment will not look right when sewn together and it will not fit well on the body.

第 8 课

服装纸样裁剪

设计师尽可能早地知道一件服装如何从二维概念变成三维的物体是非常重要的。服装纸样就是纸或硬纸的平面模板，再将纸样转移到面料上，然后裁剪和缝合。

重要的是要对体形有深刻的了解，并且知道如何将身体尺寸转移到纸样的衣片上。纸样裁剪师必须仔细地工作，确保缝制时服装衣片完美、精确地缝合在一起。

1. 原型

原型（也称为服装尺寸样本）是基础服装的二维样板模型（例如贴身上衣或贴体裙子），在其基础上变化成更加复杂的设计。原型采用的尺寸来自尺寸表格或真实人体，没有任何风格线或者缝份。但是原型必须包括人体活动和舒适的基本量，例如，上身贴体原型的松量就没有外衣的上身原型松量多。贴合上身原型要增加省道，形成腰和胸部形状，而宽松服装就不需要这些省道。

2. 纸样

纸样是由设计草图和原型发展而来。设计师或者样板师在原型上增加款式线、垂褶、褶裥、口袋和其他修改，设计一个最初的纸样。

最终纸样有一系列不同形状的衣片纸样，将它们转移到面料上，裁剪和缝合，得到一件三维的服装。

每个样片上要打剪口或标记点，和其他样片连接点相对应，不管是谁制作服装，都能将衣片精确地缝合到一起。衣片必须精确地缝合在一起，否则，当服装缝制好以后看上去不对劲，也不能很好地贴合身体。

专业词汇

1. transfer *vt.* 转移
2. pattern cutter 样板师
3. sloper *n.* 原型
4. bodice *n.* 上衣身
5. size chart 尺寸表
6. ease *n.* 松量
7. draft *vt.* 画纸样
8. overcoat *n.* 大衣
9. pleat *n.* 褶裥
10. adjustment *n.* 调整
11. trace *vt.* 拓印
12. notch *n.* 刻痕
13. point *n.* 点

通用词汇

① accurate *adj.* 精确的
② elaborate *adj.* 精心制作的

When the block modification is finished, seam allowance is added to the pattern. To perfect a pattern, a toile (a garment made out of a cheap fabric such as calico) is made and fitted on to a live fitting model. Adjustments can be made on the toile before being transferred to the pattern.

3. Pattern cutting

Like all craft skills, pattern cutting can at first seem difficult and intimidating. But with a basic understanding of the rules to be followed (and broken!) the aspiring designer will soon learn interesting, challenging and creative approaches to pattern cutting. To draw the right style line in the correct position on a garment takes experience and practice. Designers who have been cutting patterns for twenty years can still learn something new—the process of learning never stops. This makes creative pattern cutting a fascinating process.

4. Samples[14]

A sample is the first version of a garment made in real fabric. It is this garment that goes on the catwalk or into a press/showroom[15]. Samples are produced for womenswear in sizes 8-10 to fit the models.

Once the sale book is closed, the samples are stored in the company's archive[16]. Some samples of past collections are taken out by designers for photo shoots, events such as premieres[3] and for reference or possible inspiration for future collections.

5. How the measurements relate to the block

Whether taking individual measurements or using a size chart, the main measurements (bust girth[17], waist girth, waist-to-hip length and hip girth) will give a good indication of the body shape the design is intended to fit. Secondary measurements may also be taken from an individual or from a size chart.

Darts can be used to control excess[18] fabric and to create shape on a garment when stitched[19] together.

6. How to read a design drawing

This is the point at which pattern cutting becomes much more creative and exciting. Once the design has been completed, the process of breathing life into a flat design drawing in order to achieve an actual garment can begin. To be able to achieve a beautiful garment shape takes time and experience. Remember nothing ever happens without practising your skills—don't be disheartened[4] if it doesn't work first time round. All outstanding fashion designers and creative pattern cutters have worked for years to perfect their skills.

当原型修改完成后，纸样上要添加缝份。为了得到完美的纸样，需要用坯布（用便宜的面料例如白坯布）制作服装，再到人体上试穿。试穿后，先在坯布上进行修改，再转移到纸样上。

3. 纸样裁剪

像所有工艺技能那样，起初纸样裁剪似乎很困难，很令人畏难。但是，当理解（打破）基本规则以后，有追求的设计师将很快学会用有趣的、挑战性的、创造性的方法去进行纸样裁剪。要想在服装的正确位置上放置准确的风格线需要经验和实践。已有20年裁剪经验的设计师仍然需要学习新知识，学习过程从未停止，创新样板就成为令人着迷的过程。

4. 样衣

样衣就是用真实面料制作的第一件服装样品。正是这件样衣要拿到T台、媒体和展示厅展示。女装样衣以8～10型号制作。

一旦销售订单结束后，样衣被放进公司的档案。一些过去收藏的样衣被设计师拿出来拍照、或参加一些首映式等活动、或作为未来作品的参考资料和灵感来源。

5. 如何测量原型的相关尺寸

无论是从个体量取尺寸，还是使用尺寸表的尺寸，一些主要尺寸（胸围、腰围、腰臀的长度和臀围），能很好地表明身体体型，有目的地设计贴体的服装。其他尺寸可能来源于个体或尺寸表。

省道可以用来控制多余的面料，当面料缝合起来时，可以设计服装的廓形。

6. 如何阅读设计画稿

从这点上看，纸样裁剪有很大的创造性和激动性。一旦设计完成后，将平面设计画稿注入生气的过程就开始了，直到变成一件真实服装。为了获得一件漂亮样式的服装，需要耗费时间和经验。记住没有实践技巧，任何事情不会发生——如果第一次作品不完美，不要灰心。所有优秀的时装设计师和创意纸样裁剪师都是工作了很多年以后才使他们的技能得到完善。

专业词汇

14. sample *n.* 样衣
15. showroom *n.* 陈列室
16. archive *n.* 档案
17. girth *n.* 周长
18. excess *n.* 多余量
19. stitch *n.* 缝迹线

通用词汇

③ premiere *n.* 首次公演
④ disheartened *adj.* 沮丧的

7. Translating drawing to block

The translation of a design drawing to pattern requires an eye trained for proportions. Most design drawings are sketched on a figure with distorted proportions. The legs and neck are too long and the figure too slender[20]. These sketches are often inspiring and wonderful to look at but unfortunately give a false image of the human body and it is a key task of the pattern cutter to address this.

8. How to mark the block

It is essential when cutting a block or a pattern that the correct information is supplied. A bodice block, for example, has to show the horizontal lines of the bust, waist and hiplines[21]. Parts of the block such as the waist and bust points should be notched or punch[22] marked (holes and notches indicate where the separate pieces of fabric will be attached[23] to one another) and the grain line[24] must be indicated. This will clearly show the position in which the pattern should be placed on the fabric. Additional information must be written clearly in the centre of the block, including whether it is a front or back piece, a tight-or loose-fitted bodice block and the sample size, preferably with the measurements and any allowances to be made when constructing the block.

Once the pattern has been constructed the seam allowance can be added. Seam allowance can vary in size from a narrow 0.5cm for a neckline[25] (to avoid having to clip or trim the seam) to 2.5 cm in the centre back of trousers[26] (to be able to let some out if the waist gets too tight). Seams that are to be joined together should always be the same width. Mark the width of the seam allowance on the block.

Usually, the block ends up being divided into further pattern pieces. At this point, therefore, the information should be reconsidered accordingly, except the grain line and front or back information, which is always transferred to the new pieces.

9. Dart manipulation

Darts control excess fabric to create shape on a garment. They can be stitched together end to end or to a zero point also known as the pivotal point (such as the bust point). Dart manipulation is the most creative and flexible part of pattern cutting. The possibilities are endless and the designer's imagination is the only limitation. Darts can be turned into pleats, gathers[27] or style lines. Their positioning on the body is very important; not only do these techniques create fit, shape and volume, they also change the style and design of the garment.

10. Slash[28] and spread[29]

This method is used to add extra volume and flare[30]. The technique involves creating slash lines that reach from one end of the pattern to the other, sometimes ending on a pivotal point like a dart ending. These slash lines will then be opened up for added volume and flare.

7. 将设计稿转变成原型

将一种设计画稿转变成纸样需要有对比例训练有素的眼光。大多数设计画稿上的人像比例都是变形的。腿部和颈部太长,人像太纤细。这些草图通常看上去令人激动和完美,但不幸的是提供了虚假的人体形象。对于纸样裁剪师来说,必须纠正这些错觉的东西。

8. 如何标记原型

当裁剪原型或者纸样时,提供正确信息是很重要的。例如,上身原型必须显示胸部、腰部和臀部的水平线。原型的某些部位,例如腰线和胸点,要有凹痕或打孔的标记(孔和凹痕表明衣片在这里对齐缝合)。必须表示纱线的方向,这将清晰地表示纸样放置在面料上的位置。其他信息必须清楚地写在原型的中央位置,包括前衣片或后衣片,贴体或宽松的上身原型和样板尺寸,更周到的是标上尺寸和制作原型时的放松量。

一旦纸样完成后,就要加放缝份。不同部位缝份的量不同,在颈部窄一些,约0.5厘米(避免打剪口或修剪缝份),裤子的后中线需要2.5厘米(如果裤腰太紧,可以放出来一些)。缝制的时候缝份要一样宽,在原型上标记缝份的量。

在原型基础上分割发展的纸样,信息要作相应调整,除了丝缕和前后衣片信息,所有其他的信息都要转移到新的纸样上。

9. 操作省道

省道能控制服装上多余的面料,从而创造形。它们可以从一端缝制到另一端,或者从一端到零点,也叫做关键点(例如胸点)。省道的操作是纸样裁剪上最具创造性和灵活性的。省道有无穷多可能性,而设计师的想象极其有限。省道可以变成褶裥、抽褶或风格线,这些手法应用在身体的何处很重要,可以设计贴体、塑造形状和体积,也能改变服装的样式和设计。

10. 剪切和展开

这种方法用来增加额外和展开的量。这种技术包括设计一条剪切线,从纸样的一端到另一端,有时在一个关键点结束,像省道的端点。然后拉开这些剪切线,增加量和展开。

专业词汇

20. slender *adj.* 苗条
21. hipline *n.* 臀围线
22. punch *vt.* 打孔
23. attach *vt.* 附着
24. grain line 丝缕线
25. neckline *n.* 领围线
26. trousers *n.* 裤子
27. gather *n.* 碎褶
28. slash *vt.* 切口
29. spread *vt.* 展开
30. flare *vt.* 张开

11. Sleeves[31]

Sleeve construction is a very special part of pattern cutting. Sleeves can be part of the bodice (laid-on sleeve) or set into an armhole (set-in[32] sleeve). Without any other design features added, a garment can look outstanding by simply creating an interesting sleeve design. The most basic sleeve block is the one-piece[33] (set-in) sleeve. Different sleeve blocks can be developed from the one-piece block, such as the two-piece[34] sleeve and laid-on sleeves, including raglan[35], kimono/batwing[36] and dolman[37] designs.

When constructing a set-in sleeve, the measurement of the armhole is essential. Therefore, the bodice front and back are constructed first and once the measurement of the armhole is established, ease is added according to the type of block (jacket block, fitted bodice block and so on). Ease is added to a pattern to allow for extra comfort or movement.

As well as allowing the sleeve to sit comfortably in the armhole, ease will also affect the fit and silhouette of a garment. Ease is distributed between the front notch and the double back notch of the sleeve. In some set-in sleeve designs, the ease is taken across the shoulder to achieve a round appearance over the shoulder point. A sleeve is sitting comfortably in the armhole when it aligns exactly with, or is set slightly in front of, the side seam of the bodice.

There are differences between one-piece and two-piece sleeves, the major one being the amount of seams that are used. A one-piece sleeve has only one seam placed under the arm at the side seam position. Therefore, the seam cannot be seen when the arm is relaxed. The two-piece sleeve has two seams; one is placed at the back, running from the position of the back double notch down to the wrist, past the elbow. The second seam is moved a little to the front, from under the arm side seam position (still not visible from the front). The look of a two-piece sleeve is more shapely and it has a slight bend[38] to the front. As such, it is possible to get a closer fit with a two-piece sleeve because of its extra seam. One-piece sleeves are used for a more casual look, whereas two-piece sleeves are mostly seen on garments such as tailored jackets or coats[39].

The laid-on sleeve is part of the bodice. Once constructed, either a part of the armhole remains or there is no armhole at all. A laid-on sleeve is most commonly constructed by separating the one-piece sleeve through the shoulder notch straight[40] down to the wristline[41] to gain a front piece and a back piece. The next step is to align the front piece of the sleeve with the bodice's front shoulder and the back sleeve with the bodice's back shoulder. From this point onwards several styles can be developed, such as batwing or kimono, raglan, gusset and dolman sleeves. The sleeve can be laid on at variant angles – the greater the angle, the more excess fabric and therefore a greater range of arm movement.

12. Collars

The collar is a versatile design feature that will enhance the style of a garment. It is attached to the neckline of the garment and allows the size and shape of the neckline to vary. Collars come in all shapes and sizes and the most common are the stand-up[42] /mandarin[43], shirt, flat, sailor[44] and lapel

11. 袖子

袖子结构是纸样裁剪中相当特别的部分。袖子可以是上衣身的一部分（套袖）或者安装在袖窿里（装袖）。一件服装不需要增加任何其他特征，只创造一个有趣的袖子，就能使服装看上去非同一般。最基本的袖子原型是一片袖（装袖）。不同袖子原型可以从一片袖发展而来，包括两片袖和连肩袖，例如和服袖和蝙蝠袖。

装袖结构、袖窿尺寸很重要。因此，设计衣身的前衣片和后衣片袖窿结构，然后测量袖窿，要根据原型的不同类型增加放松量（夹克的原型，贴体的上衣身原型等等）。加放的松量保证穿着的舒适性和运动量。

袖子放松量不仅使袖子充裕地安装在袖窿里，还影响服装的贴体性和廓型。放松量均匀分布在袖子前记号缺口到后记号缺口之间。一些装袖设计中，袖子放松量安排在肩部，使肩部有圆润的效果。如果袖子安装得准确，袖子充裕地安装在袖窿里，或者从上身侧缝处略向前倾。

一片袖和两片袖有很多不同，主要不同之处在于它们的分割缝的数量不同。一片袖只有一条缝，在手臂下方侧缝的位置。因此，当手臂下垂时看不到缝。两片袖有两条缝，一条缝在后面，从衣片袖窿的缺口标记处向下通过肘部到手腕。第二条缝从手臂下方侧缝的位置稍稍向前偏移（从前面看不到缝）。两片袖子的样式更加有形，少许向前弯曲。正是多了一条缝，两片袖有可能更加贴体。一片袖多用于休闲服装，而两片袖大多用于合体的夹克或外套。

套肩袖是上衣身的一部分。它的构造要么保留一部分袖窿，要么没有袖窿。最普通的套肩袖沿肩线到手腕有一条分割线，分成前袖片和后袖片。接下来就是将前袖片与上衣身前片的肩缝合，后袖片与上衣身后片的肩缝合。依此进一步发展为多种样式，例如蝙蝠袖或和服袖、套袖、三角插片袖、肩袖和蝙蝠袖。袖子安装的角度有多种——角度越大，面料的冗余越多，因此，手臂的运动范围也越大。

12. 领子

领子可以设计成多种风格，能够增强服装的风格。它与服装的领线缝合在一起，领线的尺寸和形状可以变化。领子有各种形状和大小，最常见的领子结构是站领或中式领、衬衫领、

专业词汇

31. sleeve n. 袖子
32. set-in 装袖
33. one-piece 一片
34. two-piece 两片
35. raglan n. 插肩袖
36. batwing 蝙蝠袖
37. dolman n. 斗篷
38. bend n. 弯曲
39. coat n. 外套
40. straight adj. 直的
41. wristline n. 腕围线
42. stand-up 立领
43. mandarin n. 中式领
44. sailor n. 水手

collar[45] constructions.

Collars can be constructed in three basic ways. The first method is a right-angle[46] construction, used for stand-up collars, shirt collars[47] and small flat collars[48] such as Peter Pan and Eton collars. Secondly by joining the shoulders of the front and back bodice together to construct the collar directly on top of the bodice block. This technique is used to construct sailor collars and bigger versions of flat collars. Finally, the lapel construction, which is extended from the centre front, from the breaking point toward the shoulder. By extending the break/roll line[49] a collar construction can be added. A version of this is the shawl collar, where the collar extends from the fabric of the garment on to the lapel without being sewn on.

摊领、水手领和驳头领。

领子有三种基本构成方法。第一种方法是直角结构，有站领、衬衫领和小平领，例如彼得潘领和伊顿领。第二种方法是将前后衣片的肩部缝合起来，然后直接将领子安装在原型衣身的上端。水手领和大型平坦的领子就是这种构造。第三种领子是驳头领结构，沿前中心线延伸，从翻折点向肩部翻折。通过延伸折线或翻折线，增加领子结构。其中之一是披肩领，领子从服装的衣片向外延伸到驳头，没有缝合线。

专业词汇

45. lapel collar 驳头领
46. right-angle 直角
47. shirt collar 衬衫领
48. flat collar 平翻领
49. roll line 翻领线

Passage 9

Draping

Draping[1] method is improved by the industrial development through 1800's to twentieth century. The invention of tape measure[2], invention of sewing machine in 1851, in 1863 of the development of paper patterns and in 1911 the production of muslin padded mannequins are the very important developments that occurred in the fashion industry. By the effect of these improvements couture designers started to use three-dimensional models to drape the fabric and check[3] the relation between the design and the body.

1. Draping and flat pattern making

There are two main approaches that can be taken when converting a design into a 3D object for the first time. The approach that is chosen is very much decided by the design that needs to be achieved, and the particular working styles of the designer and/or the pattern maker.

The first approach is by flat pattern making, whereby a pattern maker will look at the designers sketch and will select a basic pattern block from which to build the design, making alterations to a basic jacket pattern for instance and adding fullness[4], moving seamlines, changing necklines etc., until the pattern resembles the design. Then this pattern can be made up in a test fabric and a first fitting can be done on a dress makers dummy[5] or mannequin or on what is known as a fit model—a live model with the exact body measurements of the pattern. After these changes will be made, the pattern will be altered and a new toile recut until the designer and pattern maker are happy that the design is correct. This approach is predominantly based around flat pattern making, and is called "flat" because it is predominantly done to the 2D pattern on the pattern making table using existing patterns and measurements.

The second approach is very different and is much more about what the fabric wants to do, or what the fabric will allow you to do. This approach is known as "drape" or "working on the stand". Fashion draping is an important part of fashion design. Draping for fashion design is the process of positioning and pinning fabric on a dress form to develop the structure of a garment design. A garment can be draped using a design sketch as a basis, or a fashion designer can play with the way fabric falls to create new designs at the start of the apparel design process. A designer or pattern maker will begin by draping basic fabric, such as calico[6], onto a mannequin and working much more like a sculptor the fabric is smoothed, creased[7], spliced[8] and pinned until the desired shape is achieved. Once the

第 9 课

服装立体裁剪

从 1800 年代到 20 世纪的工业发展，立体裁剪技术得到了改进。1851 年卷尺和缝纫机的发明、1863 年纸样的发展、1911 年衬垫的平纹细布人体模型产品的出现，它们都对时装行业的发展起到了重要作用。正是由于这些方面的改进，高级时装设计师开始使用三维模型立体裁剪面料，检验设计和身体之间的关系。

1. 立体裁剪和平面纸样制作

将一种最新设计转变为三维物体，主要有两种方法可以采用。究竟采用哪种方法，一种是由设计决定，即需要取得什么样的效果；另一种是由设计师或样板师独特的工作方式决定。

第一种方法是平面纸样制作，样板师根据设计师的草图，选择一种基本原型样板，以它为基础，根据设计再做适当修改。例如，一件夹克基本的样板，可以增加丰满度，去除接合线，改变领线等等，直到样板与设计相似。然后，这种样板采用试验的面料，第一次在人台或人体模型试穿，或者模特试穿——一种真人模特进行试穿，其身材要与样板尺寸比较吻合。试穿之后将做一些改动，在纸样上做修改，重新用坯布裁剪，直到设计师或样板师认为正确体现设计为止。这种方法采用平面样板制作，称为"平面"是因为二维样板根据已有样板和尺寸在样板桌上完成。

第二种方法截然不同，更多关注希望面料能做什么，或面料允许你做什么。这种方法称为"立体裁剪"或"在台子上工作"。时装立裁是时装设计的重要部分。立体裁剪式的时装设计，就是将面料定位在人台上，逐步发展服装结构。一件服装可以基于设计草图，通过立体裁剪完成；或者时装设计师可以在服装设计的开始阶段，在人台上摆弄面料产生新的设计。设计师或样板师首先采用白棉布等基本面料，在人台上进行立体裁剪，很像雕塑家，将面料捋平、弄褶、接缝和别合，直到取得希望的形状，然后在面料上标上铅笔线，打剪口，注释

专业词汇

1. draping *v.* 立体裁剪
2. tape measure 卷尺
3. check *vt.* 检查
4. fullness *n.* 宽松度
5. dummy *n.* 人体模型
6. calico *n.* 白棉布
7. crease *vt.* 折缝
8. splice *vt.* 拼接

garment is roughly the right shape, the fabric can then be marked with pen lines and notch marks and annotations[9] of what piece is which, so that it can then be removed from the mannequin and flattened[10] out without later confusion. It is from this drape that a first pattern can then be traced and the lines and measurements smoothed[11] and checked, before a new first toile is sewn and ready for a fitting on a fit model.

The two processes can both be used side by side, neither is more correct, and there are times when one will be more effective than the other depending on the design and personal preference. At their most basic levels of distinction flat pattern making will be quicker for designs which are closer to existing pattern blocks, for designs which are closer to the body or more graphic in nature or for those pattern makers who feel more comfortable working in this method. For designs which are more organic in nature, have more volume or more flowing use of fabric then drape is going to be an easier choice, especially for those who like to see and experiment with the design straight up as they go along, rather than committing to one pattern straight off.

There are also times when designers will work directly on the stand using real fabrics and to be able to apply embellishment or build up texture while the design is holding the shape of the human body. But in most cases the designer will want to be able to reproduce the design and will need to have a flat pattern for production at some stage of the process.

If you've ever tried to work on the stand then you will know that the fabric will start to tell you what it will and will not do. There is a certain sensitivity to working in this way, of allowing the fabric to fold[12] just so, or be caught in such a way.

2. Why should fashion designers learn how to drape?

While the majority of companies in the fashion industry no longer use draping as part of the design process, draping is a key skill which allows apparel designers to understand what creates a great fit and how to achieve it. If a garment sample fits poorly, a designer who is familiar with how darts and seams give shape to garments can spot what is creating the fit issue and advise the factory how to correct the problem.

However, the art of draping isn't completely lost; in high fashion, couture fashion houses, evening, and lingerie companies most garments are created through draping. When draping a garment, the designer can immediately see what her apparel design will look like on the body, and immediately correct any fit or design problems before putting anything down on paper. In addition, some apparel designs are just impossible to make via flat patternmaking and need to be draped first. And some fabrics need to be experimented with on a dress form to see how they behave.

While draping for apparel design may seem like a daunting and tedious approach to creating patterns, it's actually one of the more creative parts of the fashion design process. The designer's medium is fabric, which he or she manipulates[13] with skillful hands in the creation of simple to complex designs. Draping has many attributes. Draping allows the designer to evaluate the drape at each step to assure that line and balance are in harmony[①] with design. A misplaced styleline can be

衣片，以免后来弄混淆，再从人台上取下来放平。正是这种立体裁剪得到了第一个样板，然后复制印迹和线条，用尺将线条画圆顺并检查，再进行缝制，就得到第一件样衣，然后在人体上试穿。

这两种方法可以一并使用，没有谁对谁错。何时采用何种方法更有效，取决于设计和个人的偏好。它们最基本的区别是，如果设计与已有原型样板比较接近，或比较接近身体，或更加图解化，或样板师觉得使用这种方法更加舒适，那么平面制作样板的方法更快；而设计比较复杂、面料更具立体感或悬垂性，特别是有些人喜欢经过实验，看到设计变成了现实，而不是变成非真实的样板，选择立体裁剪更加便利。

有些时候，设计师直接使用真实面料在人台上立体裁剪和修饰；或者直接在人体设计出有风格特征的款式。但是很多情形下，设计师想在设计的基础上再设计时，就需要在此时有平面纸样。

如果你曾经尝试过在人台上立体裁剪，面料将告诉你它能和不能做什么。这种方式很灵活，知道面料如何折叠或把控。

2. 为何时装设计师要学习立体裁剪？

在时装行业，很多大公司不再使用立体裁剪作为设计过程的一部分，立体裁剪是一项重要的技能，它使服装设计师知道什么是完美合体的设计和如何取得合体的效果。如果一件服装样品很不合体，而且设计师知道如何使用省道和分割塑造服装的形，那么就能知道在哪个部位解决合身问题，建议工厂如何修改这些问题。

然而，立体裁剪并没有完全失去它的艺术性；在高级时装中，高级时装屋、晚装和贴身内衣公司的大多数服装都是采用立体裁剪的方法设计的。立体裁剪时，设计师能够立刻看到她设计的服装穿在人体的样式，并纠正不合体的地方，或纠正所有设计方面的问题，然后再将衣片平展在纸上。此外，一些服装设计通过平面裁剪不可能实现，需要首先采用立体裁剪。一些面料需要在人台上进行试验，了解它们的特性。

用立体裁剪将服装设计转变成纸样，这种方法似乎令人怯步和乏味，但确实是时装设计过程中最具创造性的一部分。设计师的媒介是面料，他或她用一双灵巧的手处理从简单到复杂的设计。立体裁剪有很多特征，使设计师能够在每一步骤评估面料的状况，保证线和平衡

专业词汇

9. annotation *n*. 注释　　11. smooth *vt*. 使平滑　　13. manipulate *vt*. 操作
10. flatten *vt*. 弄平　　　12. fold *vt*. 折叠

通用词汇

① harmony *n*. 协调

easily changed for greater appeal. The designer can manipulate the fabric, deciding where to place darts, tucks[14], and other design elements, and adding fullness for gathers, flares, or diagonal[15] until he or she is visually satisfied. These eliminate the need to guess, as other pattern-making systems may require. There are designs that can be developed using any pattern-making method, but designs require only draping to control the look and esthetics[②] of the completed garment. The designer also has the advantage of being an eyewitness to and a participant in the evolving shapes, from a lifeless piece of muslin to a perfect replica of the design in three-dimensional form. Creating designs by draping may take a bit longer, but the experienced designer has learned to compete and is willing to combine the skills of draping with other methods to produce garments more quickly. Draping has flexibilities that are not offered by any other system, giving draping the greater advantage. Playing with the way fabric folds and hangs[16] on the body is a fun way to create new fashion designs that you wouldn't have thought of sitting in front of a sketchbook!

3. Fabric characteristics and term

Fabric are a powerful medium in the hands of the designer whose creativity is stimulated by the array of colors, boldness[17] of prints[18], and innovative[19] textures that are offered each season. Knowledge of the characteristics of fabrics and the distinctions among them enables designer to select the most appropriate fabric for the purpose of the garment.

Fabrics are classified according to quality, structure (weave[20], knitted[21], fused[22], nonwoven-fused, or plain[23]), texture, weight, and hand (how fabric feels to the touch[24]). They are contrasted by being either crisp[25] or soft[26], thick[27] or thin[28], heavyweight or lightweight, loosely or firmly woven, flat or textured, silky or rough, transparent[29] or opaque[30], and sleazy[31] or luxurious. To evaluate the quality of a fabric before purchasing, the designer should touch and feel the fabric, observe how well it drapes by holding and raising one end to allow the folds to fall naturally, and reaction of the fibers after being crushed[32] in the hand. There is much to learn about fabrics. In addition to reading books devoted to textiles, you should collect swatches when buying fabrics and catalogue them by width and content. In this way, you can create a personal reference library.

1) Muslin

Fashion draping and fitting are usually done with muslin to resolve any design and fitting issues of a garment before cutting the pattern in real fabric. In draping the drapery of the chosen fabric is very important because it affects the finished look of garment. The muslin should be chosen carefully according to the real fabric used in design. Garments made of woven[33] fabrics should be draped with muslin or cheaper fabric where weft[34] and warp[35] yarns can be easily seen. The quality and the hand of the muslin should represent the fabric that is used in design. Garments made of knit should be draped by using inexpensive knitted fabric which has the same stretch[36] with the real fabric used in design.

通用词汇

② esthetic *adj.* 审美的

与设计取得和谐，错误的样式线条能轻易地改变服装的整体形象。设计师能够操作面料，决定在哪里放置省道、折叠和其他的设计元素，用抽褶、展开或斜向面料增加丰满度，直到他或她满意为止，而不用像其他裁剪方法那样需要猜测。有很多设计可以任意采用某种纸样裁剪的制作方法，但是设计需要通过立体裁剪控制成品服装的样式和美感，有利于设计师目睹和参与形态的演变过程，从毫无生气的白坯布到完美的三维设计形态。通过立体裁剪方法进行的设计可能需要的时间长一点，但有经验的设计师有能力并且乐意结合其他方法使立体裁剪更加快速。立体裁剪具有其他任何裁剪方法没有的灵活性，这就是立体裁剪的优势。将面料在人体上折叠和悬挂就能创造崭新的时尚设计，而不必坐下来面对速写本，真是一件很有趣的事情。

3. 面料特征和术语

面料是设计师进行设计的有效媒介，每一季提供的各种色彩、大胆印花和创新肌理的面料，都能激起设计师的创意。熟悉面料的特征和它们之间的差别能使设计师选择最适合服装用途的面料。

面料根据品质、结构（梭织、针织、热熔、非织造热熔或平纹），肌理、重量和手感（手的触感）分类。它们的对比是：脆硬或柔软、厚或薄、重或轻、松织或紧纺、平纹或肌理、绸感或糙感、透明或不透明、便宜或昂贵。在购买前设计师要评估面料的品质，触摸和感受面料；将面料折叠起来，然后拎起面料一角，松开面料，让它自然下垂了解它的悬垂性；将面料在手中挤压，观察纤维的反应，有很多关于面料的知识需要学习。除了阅读面料书籍，在你购买面料时，应该收集面料小样，并且根据它们的宽度和成分进行分类。通过这种方法，你可以创造一个个人资料图书馆。

1）平纹细布

时装立体裁剪通常使用平纹细布，解决服装上任何设计和合身方面的问题，然后再在真实的面料上裁剪衣片。在立体裁剪中，选择立体裁剪用的面料很重要，它影响服装成品的面貌。要根据设计中使用的真实面料仔细挑选平纹细布。梭织面料服装，应该使用平纹细布或便宜面料进行立体裁剪，纬向和经向要很清晰。平纹细布的质量和手感必须与设计中使用的面料相似。针织面料服装的立体裁剪要使用便宜的针织面料，要和设计中的真实面料有相同的弹力。

专业词汇

14. tuck *n.* 箱型褶裥
15. diagonal *n.* 对角线
16. hang *vt.* 悬挂
17. boldness *adj.* 大胆
18. print *n.* 印花
19. innovative *adj.* 创新的
20. weave *n.* 织物
21. knit *n.* 针织

22. fuse *vi.* 热熔
23. plain *n.* 平纹
24. touch *n.* 手感
25. crisp *adj.* 挺括
26. soft *adj.* 柔软
27. thick *adj.* 厚的
28. thin *adj.* 薄的
29. transparent *adj.* 透明的

30. opaque *adj.* 不透明
31. sleazy *adj.* 质地薄的
32. crush *vt.* 挤压
33. woven *v.* 梭织
34. weft *n.* 纬纱
35. warp *n.* 经纱
36. stretch *n.* 拉伸

Muslin fabric, the most common material used for draping, is inexpensive and falls loosely over the dress form, making it easy to manipulate to create different looks. It is a plain-woven fabric made from bleached[37] or unbleached yarns in a variety of weights, including:

Lightweight muslin: It represents the drapery of natural and synthetic silk, light cotton fabrics and lingerie fabrics.

Medium weight muslin: It simulates the wool and medium weight cottons.

Course muslin: It simulates heavy weight wool and cotton fabrics.

Canvas muslin: It is used to drape denim, fur and some heavy weight fabrics.

2) Grains and lines

Selvage[38]: The narrow, firmly woven finished edge[39] on both sides of the fabric length. To release the tension[40], clip[41] along the selvage edge.

Grain: The direction in which a fiber is woven or kintted.

The weft and warp directions or the grain line refers to the orientation of yarns in woven fabrics. The weft and warp yarns are perpendicular[42] to each other in the weaving loom[43].

The warp yarns form the lengthwise grain and parallel[44] with the selvage. It is twisted[45] more tightly than the crosswise[46] grain and also called straight grain. The weft yarns form the crosswise grain woven across the fabric from selvage to selvage.

The drape of the fabric varies in different directions since the warp yarns resist[47] stretching most of the garments are cut in lengthwise[48] direction, perpendicular to the hem[49]. As the weft yarns stretch more than warp yarns, they can easily adapt the movements of the body. So the crosswise grain of the fabric is mostly used around the body. The crosswise grain of the fabric is very seldom placed vertically because the yarns will relax[50] and the garment will droop.

Any angle that falls between crosswise grain and lengthwise grain is called the bias[51]. True bias is at a 45-degree angle to lengthwise and crosswise grain. True bias has maximum give and stretch, easily confirming to the contour of the figure. Flares and cowls drape best when on true bias. The designers use bias cut to give the dresses more shape and fullness. Because of the stretch and distortion[52] of the bias, the cutting and sewing of bias cut fabric patterns need more attention.

4. Model form

For the past 140 years, forms have adapted to the whims of fashion by constantly modifying in shape and measurement to satisfy the needs of changing silhouettes. Original forms were shapeless, willow-caned models with woven mounds that were padded to individual specifications. Today's forms are partially made by hand. They are framed in metal, molded with papier-mâché, laid over with canvas, and covered in a princess garment of linen[53]. The seam lines of the cover garment set the boundaries between the front and back bodice and skirt. The waistline[54] seam defines the upper and lower torso.

A new innovation in form making creates forms that feel flesh[55] to the touch and can be penetrated with pins without harm to the form. Forms of today represent the most common

平纹细布是立体裁剪时普遍采用的面料，便宜而且能随意地覆盖在人台上，容易操作，创建不同的外观。它是一种平纹梭织面料，用漂白或不漂白纱线制造，重量有多种，包括：

轻型平纹细布：替代天然和人造丝绸、轻型棉织物和内衣面料的立体裁剪。

适中型平纹细布：替代羊毛和重量适中的棉质面料。

粗糙平纹细布：替代重型羊毛和棉织物。

帆布平纹细布：替代牛仔、裘皮和一些重型面料。

2）丝缕线

布边：面料长度方向两边窄的、坚固的织边。为了释放张力，沿着布边将它剪去。

丝缕：梭织或针织时纱线的方向。

直丝缕和横丝缕是指面料纺织时的纱线方向。纬纱和经纱在织布机上相互垂直。

经纱构成直丝缕，平行于布边，它的纱线捻度比纬纱紧。纬纱构成了横丝缕，从布边的一边到另一边。

立体裁剪的面料可以变化不同方向，因为直丝缕不易拉伸，大多数服装裁剪时都将直丝缕与底边垂直。因为横丝缕比直丝缕容易拉伸，它们能够适合身体的运动，因此，面料的直丝缕大多用来包围身体。面料很少按横丝缕方向垂直于地面放置，因为纱线会松弛，服装也将会下垂。

在直丝缕和横丝缕之间的任何角度都称为斜丝缕。正斜是指与直丝缕和横丝缕都成45度角度。正斜丝缕具有最大的弹性和拉伸力，有效地体现身体的轮廓线。波浪和垂荡形最好采用正斜丝缕裁剪。设计师使用正斜丝缕裁剪，使连衣裙更具有形和丰满性。因为斜丝缕容易拉伸和变形，裁剪和缝纫时需要特别注意。

4. 人台

在过去140年里，人台不断修改形状和尺寸，以适合时尚的奇思怪想，满足廓型不断变化的需求。最初人台是用不定型的柳条编织成模型，再根据个体的尺寸进行衬垫。今天有些人台是手工制作，其框架采用金属，模型用纸浆，再用帆布包裹，上面再覆盖一层有公主线分割的亚麻布。上面一层的服装有缝线，区分上身前后片和裙子的前后片。腰线界定了躯干的上部分和下部分。

一种新制造发明的人台在触摸时更具肉感，可以插针而不会对人台造成损害。今天，有男性、女性、孩子、成人不同群体平均尺寸的人台。人台有一个可拆卸的手臂、腿和可拆卸

专业词汇

37. bleach *vt.* 漂白
38. selvage *n.* 布边
39. finished edge 实边
40. tension *n.* 张力
41. clip *vt.* 剪掉
42. perpendicular *adj.* 垂直的
43. loom *n.* 织布机
44. parallel *adj.* 平行的
45. twist *vt.* 加捻
46. crosswise *adj.* 横的
47. resist *vt.* 抵抗
48. lengthwise *adv.* 纵长地
49. hem *n.* 底边
50. relax *vt.* 放松
51. bias 斜丝缕
52. distortion *n.* 变形
53. linen *n.* 亚麻布
54. waistline *n.* 腰围线
55. flesh *n.* 肉体

dimensions within each size group of males and females, children to adults. Forms come with detachable arms and legs, and collapsible[3] shoulders for ease of entry.

5. Basic dress foundation

Fashion designers drape garments in sections i.e.: front bodice, back bodice, front skirt, back skirt etc. and only the right side of the garment is draped, unless the apparel design is asymmetrical[56].

The simplicity of the basic dress makes it an ideal choice for the introduction to draping. The principles of draping and related draping techniques will be applied to complete the drape of the basic dress. After mastering the fundamentals of the draping process, the designer will advance with confidence to more complicated designs that build on these applications.

The dress is draped to fit the dimensions of the form, or figure and bridges hollow[4] areas between the bust, buttocks[57], and shoulder blades[58]. Ease is added for comfortable movement without the appearance of stress. The sleeves hang with relaxed arms in the perfect alignment[5]. The skirt straight from the widest part of the hips and the hem is parallel to the floor. A number of darts control the fit of the garment by taking up remaining excess where it is not wanted, such as the ends of the radiating[59] points of the bust, buttocks, and shoulder blades.

The foundation of the basic dress is related to all draped garments by the application of the draping and technical principles learned in its creation, including the relationships among line, balance, and fit. To drape a perfect garment takes time and patience. Every accomplished designer knows that hard work and perseverance[6] is the key to perfection.

The following information discusses three draping principles and techniques that apply to the process of developing designs.

1) Three draping principles

Principles of the dart excess—The dart is a fitting device[60] that controls remaining excess within the boundaries of the marked drape. The dart is identified by its wedge[61] shape, formed in the process of draping. The width depends on the amount of excess; length depends on distance from its source.

Principles of the cross-grain—The cross-grain placed on the form parallel with the floor will balance a garment and divide fullness above and below a guideline. Fall of the cross-grain creates bias and flare or lift for drapery, pleats, and gathers.

Principles of a balanced sleeve—A well-balanced sleeve aligns with or is slightly forward of the side seam of the form. A sleeve will hang out of alignment if the shoulder line and/or side seam of the form have not been adjusted. The sleeve must align with the angle of the client's arm and stance[62].

2) Three draping techniques

Moving dart excess—Excess is moved along the seam lines (boundaries) of the form while smoothing muslin around the bust, or any mound, to locations directed by the design. Excess may be in the form of a dart equivalent in creating the design.

Adding fullness—To create fullness, slash and drop the cross-grain; lift the grain for fullness to create drapery, gathers, and pleats. Use the ratio[63] rule to determine fullness needed. Do this when the

的肩，是为了穿衣方便。

5. 女装基础样板

时装设计师按照服装的不同部位进行立体裁剪，例如：前衣身、后衣身、前裙片、后裙片等等，通常仅立体裁剪服装的右半部分，如果服装设计是不对称的，则立体裁剪全部。

女装基础样板是介绍立体裁剪的理想选择。女装基础样板的立体裁剪中应用了立体裁剪的基本原理和相关技术。掌握了立体裁剪基础知识以后，设计师将更有自信地把基础知识应用到复杂的设计中。

女装基础样板的立体裁剪就是贴合人台或人体的尺寸，弥合胸部、臀部和肩胛骨区域的空隙。增加松量是为了运动的舒适性，服装不会出现拉扯。在手臂放松时袖子处于完美和顺的状态。裙子从臀部最宽部位到裙摆要直，裙摆平行于地板。一些省道能控制服装的贴合性，即通过收掉不需要的多余量，例如以胸部、臀部和肩胛骨等部位为末端点的放射状省道。

女装基础样板的立体裁剪知识可以应用到所有服装的立体裁剪中，包括线条、平衡和合身的关系。立体裁剪使服装达到完美状态需要时间和耐心。每一个有成就的设计师都知道，艰苦工作和持之以恒是获得完美服装的关键。

以下信息讨论在设计进展过程中立体裁剪应用到的三个基本原理和技术。

1）立体裁剪三个基本原理

省道余量原理——省道是一种合身设计，它能控制立裁中标记边界以内多余的量。省道形状为楔形，在立体裁剪过程中形成。其宽度依赖于多余的量，长度根据它到源头的距离。

横丝缕原理——横丝缕放在人台上时要与地面平行，使服装取得平衡，在引导线上方和下方作丰满度的分配。横丝缕下降产生斜向和波浪，横丝缕提升产生垂荡、褶裥和碎褶。

袖子平衡原理——袖子取得很好的平衡状态下，与人台侧缝一致或略微向前。如果肩线和侧缝不修正，袖子可能歪着。袖子必须与客户手臂的角度和姿态一致。

2）立体裁剪三种技术

转移省道余量——余量沿着人台缝线（边界线）移动到设计指向的部位，但是，围绕胸部或任何隆起部位的布要捋平。

增加丰满度——为了取得丰满状态，剪切和下降横丝缕，或提升横丝缕产生悬垂、碎褶

专业词汇

56. asymmetrical *adj.* 不对称
57. buttock *n.* 臀部
58. shoulder blade 肩胛骨
59. radiate *vt.* 放射
60. device *n.* 装置
61. wedge *n.* 楔形
62. stance *n.* 站姿
63. ratio *n.* 比率

通用词汇

③ collapsible *adj.* 可折叠的
④ hollow *adj.* 空的
⑤ alignment *n.* 队列
⑥ perseverance *n.* 坚持不懈

dart excess is insufficient.

Contouring—To reveal the contours of the form or figure, remove the bust bridge to allow fabric to be draped into the hollows around the bust (padded for emphasis, or unpadded). Side ease is removed for strapless gowns[64].

6. Manipulation of draping in couture design

In couture design, draping method is used very effectively. For using draping, couture designer should know the human anatomy[65], fibers and fabric. The silhouette and the fabric should be in harmony since the weight, texture, hand and drape of the fabric might not be appropriate for some silhouettes. If the designer wants to create sculpted look, crispy woven fabric should be chosen. If he works with very soft fabric, the design will follow the curves of the body and drape gently.

In a couture house after the appropriate fabrics are selected, the designer works with the fabric to see how it hangs on the lengthwise grain, crosswise grain or bias. Then the designer starts sketching for his collection and sketches are distributed to the workrooms of the house. More structured garments like jackets and coats are sent to tailoring room, night gowns, dresses and skirts which need to be draped are sent to dressmaking room. Dressmaking room is where designs are draped and sewn on a dress form temporarily. In the dress making room, muslin which has similar drape and hand with the real fabric that is going to be used in dress is chosen. Dress forms are padded according to body measurements of the individual.

Then the design chosen is draped on the dress form by a sample maker. By this way basic patterns for the garment are created, this process takes 4 to 8 hours according to the complexity of the design. These patterns are basted together for the first fit. Even though patterns are working patterns, every detail like buttonholes[66], trims are carefully applied. After the fit, the patterns are corrected; this corrections and fittings are repeated until the designer is satisfied with the look.

Sometimes while draping, the designer doesn't use any inner structure and the garment shows the lines of the body. Madeleine Vionnet' bias cut dresses and Madame Gres' dresses are the best example for this type of dresses. Sometimes designers use inner structure to give shape, and this structure is like a skeleton that holds the drapes of the garment in place. We can see this type of design philosophy in Charles James' garments.

The visual and functional properties of a garment are directly connected to the relation between body and the garment. A garment might have a good design but might not give enough freedom for movements of body. For this reason, the two dimensional patterns should follow the three dimensional form of the body and the best way of providing the perfect fit is draping.

There are several advantages of using draping in garment design. First of all since the three dimensional effect of the garments is very different than the effect on paper; a new and much innovative design can be created by draping. Since the designer works with the fabric, he can use the fabric with its maximum potential in his designs. Sometimes the drapes and folds of the fabric may be very different than it is sketched on paper.

和褶裥。当省道多余量不够时，使用比率规则可以满足丰满度的需求。

　　轮廓线——为了显示人台或人体曲线，将胸部两乳之间搭建的布条去掉，使面料覆盖到胸部周围凹陷的部位（此时为了强调胸部采用衬垫，或者胸部不衬垫）。无带裙装去掉侧缝的放松量。

6. 高级时装设计中立体裁剪的运用

　　在高级时装设计中，使用立体裁剪方法十分有效。为了使用立体裁剪，高级时装设计师必须知道人体解剖、纤维和面料的知识。廓型和面料必须协调，因为面料的重量、肌理、手感和悬垂性可能不适合一些廓型。如果设计师想创造雕塑般的形象，应该选择硬挺的梭织面料。如果他采用十分柔软的面料，设计将跟随身体曲线，微微下垂。

　　在高级时装屋，当选择了适当的面料以后，设计师就要开始审视面料，查看面料的直丝缕、横丝缕和斜丝缕的悬垂性。然后，设计师开始为他的设计画草图，再将草图送到各个车间。夹克和外套这些多结构的服装被送到男装车间；需要立体裁剪的晚装、连衣裙和裙子被送到女装车间。在女装车间，设计师在人台上进行立体裁剪和临时缝合。在立体裁剪车间里，选择与设计中使用的真实面料相类似的坯布。人台要根据个人的身体尺寸进行衬垫。

　　然后，挑选出来的设计由样衣师在人台立体裁剪。通过这种方法，得到服装的基本衣片，根据设计的难易程度，这个过程需要4～8小时。这些衣片被粗缝在一起，进行第一次试穿。即使样板是工作样板，每个细节、装饰都要仔细地加上，例如纽扣洞。在试穿之后，衣片要修改，这种修改和试穿要进行好多次，直到设计师满意这种样貌为止。

　　有时设计师使用立体裁剪时不使用任何内部结构，服装就能显示身体的线条。马德琳·维奥内的斜裁连衣裙和格瑞斯夫人设计的连衣裙是这种服装的最好事例。有时设计师使用内部结构塑造形状，这种结构就像骨骼，使服装处于正确的平衡状态。我们可以在查尔斯·詹姆斯的服装设计中看到这种结构原理。

　　一件服装的视觉和功能特征，直接与身体和服装之间关系相关。也许服装设计得很好，但是没有足够空间可以自由活动。正是这个原因，二维纸样应该遵循三维身体模型，提供完美合身的最好方法是立体裁剪。

　　在服装设计中，使用立体裁剪有很多优势，首当其冲的是服装的三维效果与纸上的效果明显不同，一种新的、十分有创意的设计可以通过立体裁剪实现。因为设计师的工作对象是面料，他可以在设计中发挥面料最大的潜在特征。有时，面料的悬垂和折叠与纸上的草图完全不同。

专业词汇

64. strapless gown 无吊带裙袍　　　65. anatomy *n.* 解剖　　　66. buttonhole *n.* 扣眼

By the help of draping a wrong fitted garment is quickly recognized and can be corrected. When the body measurements are different than the standard measurements of the flat pattern method, the patterns may not fit to the body. Since draping is done on a model that is modified according to the body measurements, the patterns will perfectly fit to the body. Draping is a procedure that affects the garment quality directly. Although draping is very expensive and time consuming when it is practiced correctly it gives the best result.

立体裁剪有助于很快地识别服装哪里不合身，并做出快速的修改。当身体尺寸与平面纸样的标准尺寸不同时，样板可能不合身。而立体裁剪是在人体上实现的，可以根据身体的尺寸做修改，样板将能很好地适合身体。立体裁剪是直接影响服装质量的一种工序，尽管立体裁剪的花费贵了一些，在不断修改时消耗时间，但是它能够取得最好的效果。

Passage 10

Pants Pattern

Pants[1] have become an important part of the modern wardrobe and may be worn for nearly every occasion. Even though the names and styles of pant-type garments change with fashion, the need for a good fit is always the same. A classic fitted pair of pants with slanted① pocket, front fly zipper, contoured waistband[2], belt[3] loops② and back dart. Center front hits the belly③button.

1. Pattern selection

The pattern size of pants is determined by the hip size, taken eight to nine inches below the waistline. This keeps pattern alterations to a minimum. Choose a pattern with a full-length leg, even though you plan to make shorter pants. The grainline cannot be properly established unless a full-length pattern is used.

2. Fabric selection

Select a firmly woven fabric. All wool or predominately④ wool fabrics are recommended, but other fabrics may be used. Do not select stiff fabric if you are making tapered pants.

Plaids[4] may be used. Avoid unbalanced plaids unless you have had experience in handling plaids. Linings[5] may be of a man-made fiber or a cotton blend⑤. Batiste[6] or acetate[7] is not usually recommended⑥.

Findings needed include: a zipper, matching thread, two hooks⑦ and eyes or a wide, flat hook and eye, one snap fastener[8] or a button and matching grosgrain[9] ribbon[10], one-inch wide and nine inches longer than your waistline measurement.

3. Accurate measurements

Pants patterns cannot be fitted on the figure. Therefore, it is important to know your measurements and understand your figure variations in order to select and prepare a pattern for making a comfortable, well-fitting garment.

专业词汇

1. pants *n.* 裤子　　　　　2. waistband *n.* 腰带　　　　　3. belt *n.* 腰带

第 10 课

裤子纸样

裤子已成为现代衣柜里的重要组成部分，几乎每一场合都可以穿。尽管裤装的名称和款式随时尚而变化，但合身需求始终是相同的。一条经典合体的长裤有斜袋、前门襟拉链、弧形裤腰、腰带环、后省和触及肚脐的前中门襟。

1. 纸样选择

裤子纸样尺寸由臀围尺寸决定，它在腰线下方 8～9 英寸，这样保持纸样最小的改动。即使你计划制作一条短裤，只有选择使用全腿长纸样才能正确地确立丝缕线。

2. 面料选择

选择结实的梭织面料。推荐所有羊毛或羊毛成分含量高的面料，但是其他的面料也可以使用。如果制作锥形裤，不要选择硬挺的面料。

可以使用格子面料，避免不均衡格子面料，除非你有经验把控它。里子可以是人造纤维或者是棉混纺。通常不建议采用细棉布、醋酸人造丝里衬。

需要的附件包括：一根拉链、匹配的线、两个钩扣或一个宽扁平钩扣、一副揿扣或一个纽扣和匹配的罗缎带，罗缎带尺寸是宽一英寸、长比腰围尺寸多九英寸。

3. 精确的尺寸

裤子纸样在人体上不可能合身，因此，知道你的尺寸和了解你的体型与纸样不同之处是重要的，目的是选择和准备一个纸样，制作舒适和合身的服装。

---

4. plaid *n.* 格子图案
5. lining *n.* 衬里
6. batiste *n.* 细棉布
7. acetate *n.* 醋酸人造丝
8. snap fastener 揿扣
9. grosgrain *n.* 罗缎
10. ribbon *n.* 缎带

通用词汇

① slant *vt.* （使）倾斜
② loop *n.* 环
③ belly *n.* 腹部
④ predominate *vi.* 占支配地位
⑤ blend *vt.* 混合
⑥ recommend *vt.* 推荐
⑦ hook *n.* 钩

You will need someone to take your measurements. Wear the undergarments[11] you would normally wear with pants. Place a fitting band or a tape around the waistline for a point of reference for all vertical measurements. It should be snug[12] enough to support the pants but loose enough to be comfortable.

Record your measurements in the first column of the chart 10.1, including the amount of ease suggested. Measure the pattern and record the measurements in the middle column. Measure the entire crotch[13] length as shown in Figure10.1. Match the inseams[14] of the back and front legs at the lowest point of the crotch. Place the tapeline on edge to measure.

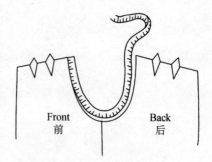

Figure10.1　Measure the entire crotch length

图 10.1　测量整个裆长

Chart 10.1　Personal and pattern of measurements

	Personal measurement	Pattern measurement	Alteration needed
Waist measurement plus 0-1" ease; add ease in garment, not waistband			
Hips at fullest part plus 2" ease____inches below waistline			
Crotch measured between the legs from center front waistline to center back waistline plus ½-1" ease for average figures, 3" for full figures. Crotch seam should be fairly snug. Take measurement when standing. You can check this by comparing your measurement with the crotch measurement of a comfortable pair of pants			
Thigh at the crotch plus 2" ease			
Knee plus 2" ease			
Calf[8] plus 2" ease			
Instep[9]			
Length of outseam (outside leg seam) from waist to ankle bone or to desired length			
Length of outseam to center of knee			

Compare the measurements in both columns to determine the amount the pattern should be enlarged or made smaller. Record the difference in the right hand column. Use a plus sign before the

你需要他人为你量尺寸。穿上经常穿在裤子里面的内衣，围绕腰线放置一条合身的带子或一根软尺，作为所有垂直方向测量的参照点。腰带应该足够紧身，得以支撑裤子，但是又要足够地宽松，确保舒适。

在表格 10.1 的第一纵列中记录包含松量的你的尺寸。测量纸样尺寸，记录在中间一列中。量取裤裆的全部长度。在裆部最低点，将后裤片和前裤片的下裆缝对齐，用软尺沿裆部边缘测量（图 10.1）。

表格 10.1　个人和纸样尺寸

	个人尺寸	纸样尺寸	修改需求
腰围尺寸增加 0～1 英寸松量；在服装上添加松量，而不是裤腰上			
在腰围线下方____英寸，臀部最丰满的部位增加 2 英寸松量			
在两腿之间，从腰围线前中心点到后中心点测量裆长，正常个子再增加 1/2～1 英寸，高个子增加 3 英寸。裆缝应该适当地紧身。测量时人体站立。你可以将测得的尺寸和一条舒适裤子的裆长做比较			
裆部大腿增加 2 英寸松量			
膝部增加 2 英寸松量			
小腿腓部位增加 2 英寸松量			
脚背			
从腰到踝骨或希望的长度，测量外缝（外侧腿缝）长度			
外缝到膝盖中部长度			

比较两列中的尺寸，决定纸样应该扩大还是变小的量，在最右一列中记录它们的差量。

专业词汇

11. undergarment *n.* 内衣　　13. crotch *n.* 裤裆　　14. inseam *n.* 内长
12. snug *adj.* 紧身的

通用词汇

⑧ calf *n.* 腿腓　　⑨ instep *n.* 脚背

number of inches it must be enlarged (+3) and a minus sign if it must be made smaller (-3).

4. Pattern preparation

Do not trim seam allowances from printed patterns until they are measured, altered, and the necessary markings have been added. Patterns may differ in the location of the lengthwise grainline, the shape of the crotch, the length of the back crotch, and the location of the darts.

Establish the lengthwise grainline in the back and front. If the garment needs to be widened, make the necessary alteration before marking the grainline. To increase the width, make the alteration between the outer dart and the side edge of both pattern pieces, parallel to the lengthwise grainline printed on the pattern.

Fold and crease each piece lengthwise, matching the seam markings on the pattern at the hem and knee. Unfold the pattern and draw a line on the crease. The creases in the pants will be pressed on the grainlines just established.

Draw the hipline at the seven-inch mark, the crotch line, and the knee line, perpendicular to the grainlines, as in Figure 10.2.

The front and back crotch (center) seams on some patterns, unfortunately, are so slanted that the finished seam gives the appearance of a dart and makes fitting difficult. Pants pull down in back when you are seated if the back crotch seam is too bias. In order to straighten the crotch seams, draw a line parallel to the corrected grainlines, starting on the seamline about an inch above the deepest curve of the crotch up to the waistline. See Figure 10.3. This makes the curve of the crotch a "U" shape and gives a more comfortable fit. Apply colored tissue paper with rubber cement[⑩] where needed.

Alter the pattern if necessary. If the garment is to be lined, the lining, which is cut after the first fitting, must be the same size and shape as the garment. If any changes are made during fitting, alter the pattern the same amount. It can be used again if your figure does not change, without having to fit each garment.

Most commercial patterns have lines indicating where pants may be lengthened or shortened. This is an easy way to enlarge a pattern. Draw two parallel lines on tissue paper, the distance between them the same amount as the pattern will be spread. Draw them a little longer than the line that will be cut and spread. Paste the cut edges of the patterns on the lines on the tissue paper with rubber cement. Keep the grainline straight.

Alterations in length of crotch. If too short, cut the pattern on the lengthen-or-shorten line and spread the two pieces the amount needed. Paste on tissue, keeping the grainlines straight and the cut edges parallel. Make the same alteration in back and front.

If too long, fold a tuck on the lengthen-or-shorten line to remove the excess amount. If the pattern is not marked, alter on the seven-inch line below the waistline.

通用词汇

⑩ rubber cement 橡胶胶水

在英寸单位数字之前添加"+"号，表示必须增大的量（+3）；添加"-"号，表示必须缩小的量（-3）。

4. 纸样准备

在测量、修改和添加必要的标记以后再修剪印刷纸样的缝份。纸样的挺缝线位置、裆的形状、后裆长度和省位可能不同。

在后片和前片上确立挺缝线。如果服装需要增宽，在标记挺缝之前作必要的修改。要想增加宽度，在前后片上，在外侧省和侧缝之间作修改，平行于印刷纸样的挺缝线。

折叠前后片的挺缝线，并留有折痕，对齐纸样上底边和膝部的缝合标记。展开纸样，在折痕上画线。在裤子上的折横要被熨烫成定型的挺缝线。

标记七英寸处的臀围线、上裆线、膝盖线，它们都垂直于挺缝线。（图10.2）

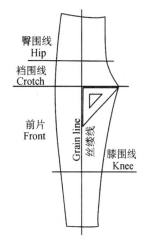

Figure 10.2　Drawing the hipline, crotch and knee line

图 10.2　画臀围线、上裆线和膝围线

不幸的是，在一些纸样上，前和后裆（中心）缝很斜，使得缝合以后的缝像省的形状，合体性很困难。如果后裆缝太斜，当你坐下来时，裤子后面往下拉。为了使裆缝取直，从裆弧线最低点向上一英寸到腰围线，画一条线平行于已经修正了的挺缝线。（图10.3）这样使裆部弧线成"U"形，更加舒适合体。在需要的地方使用带有橡胶胶水的有色棉纸。

如果需要的话，修改纸样。如果裤子有里子的话，那么第一次试穿之后裁剪，它与裤子纸样尺寸和形状相同。在试穿之后如果作任何修改，里子也修改相同的量。如果你的体型没有变化，纸样能够再次使用，不要再试穿每一条裤子。

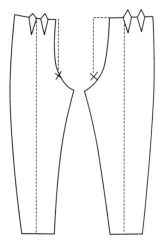

Figure10.3　Correcting the front and back crotch seams

图 10.3　纠正前后裆缝

大多数商业纸样都有一些线条，表明裤子在哪里加长或缩短，这是一种加大纸样的简单方式。在棉纸上画两条平行线，之间的距离是纸样上将要展开的量。所画的线条要比将要剪切和展开的线条略长一些。将纸样剪切的边缘用橡胶胶水沿绵纸上的线粘贴，保持丝缕线成直线。

修改裆的长度。如果裆太短，剪开纸样的伸缩线（lengthen-or-shorten），将两片展开需要的量。粘贴在棉纸上，保持丝缕线成直线，两剪切边平行。在后片和前片上作相同的变化。

如果太长，在伸缩线（lengthen-or-shorten）折叠，去掉多余的量。如果纸样上没有标记，在腰围线下方七英寸处修改。

Alterations in length of leg. Lengthen or shorten, as needed, midway between the crotch line and the knee line or the knee line and the hemline.

Measure down from the waist at the side to determine the leg length desired. Mark the pattern. Allow 1 ½ to 2 inches for a hem. If making shorts, add ½ inch to the inseam length to give the appearance of an even hemline when the garment is finished.

Add one inch seam allowance, for fitting purposes, at the waistline and outseam and inseam (inside leg seam). Apply tissue paper where needed. Trim pattern edges after all seam allowances have been drawn. See Figure 10.4.

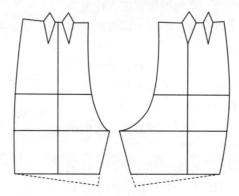

Figure10.4　Adding ½ inch to the inseam length
图 10.4　下裆缝上增加½ 英寸

5. Designing your own pants

Pants legs with varying degrees of fullness may be designed from a basic pants pattern.

1) Bell bottom trousers

Determine the amount of fullness you want at the hemline. Measure the hemline of both pattern pieces. Subtract[①] this number from the amount of fullness desired. Divide the remainder by four to determine the amount that must be added to each seam at the hemline.

For example:

$$\begin{array}{r} \text{Fullness desired} = 25 \text{ inches} \\ -\text{Pattern measurement} = 17 \text{ inches} \\ \hline = 8 \text{ inches} \end{array}$$

(8 divided by 4 equals 2 inches to be added to each seam) Draw lines from the hemline up to the knee line as shown in Figure 10.5.

2) Hostess pants

Widen the hemline on both sides of each pattern piece a minimum of four inches. Draw lines from the extended hemline to the hipline and the point of the crotch as shown in Figure 10.6.

If a shorter length is desired, extend the hemline at that level and draw up to the hipline and point of crotch.

修改腿部长度。在上裆线和膝围线或膝围线和底边线的中间位置上拉长或缩短需要的量。从腰部侧缝向下量,决定腿部需要的长度。标记纸样。底边可以是 3/2 到 2 英寸宽。如果制作短裤,在下裆缝增加 1/2 英寸,保证成品裤子的底边呈现均衡状态。

为了合身目的,在腰围线、侧缝和下裆缝(腿内侧缝)增加一英寸缝份。在需要的地方应用棉纸。在所有缝份画完后,修整纸样边缘(图 10.4)。

5. 设计你自己的裤子

从基础裤子纸样可以设计不同宽大程度裤管的裤子。

1)裤子底部呈铃形

决定你想要的底边宽大的量。测量前后片的底边,从你想要的宽大的量中扣除这个量。将剩余的量分为四等份,决定底边每一侧缝增加的量。

例如:

$$\begin{aligned} 希望的量 &= 25 \text{ 英寸} \\ -纸样尺寸 &= 17 \text{ 英寸} \\ \hline &= 8 \text{ 英寸} \end{aligned}$$

(8 除以 4 等于 2,添加到每一缝边上)从底边到膝围线画线(图 10.5)。

2)女主人裤

前后片底边每侧加宽至少 4 英寸。从延长的底边线到臀围线和裆端点连线(图 10.6)。

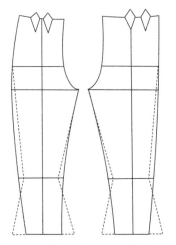

Figure10.5 Drawing the bell bottom lines
图 10.5 画铃形底边线

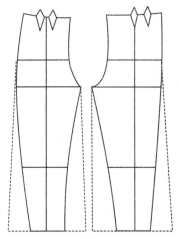

Figure10.6 Widening the hemline on both sides
图 10.6 在两侧加宽底边线

如果想长度短一些,那就在需要的长度那里延伸底边,然后向上与臀围线和裆点连接。

通用词汇

⑪ subtract *vt.* 减去

6. Pattern layout[12] and cutting

Refer to the pattern instruction sheet for layouts. One-way designs and napped[13] fabrics must be cut with the pattern pieces going in the same direction. If you are using plaid fabric, decide which lengthwise lines you want at the outseams. Sometimes the tapering necessary to fit the waist and legs causes the lines to form a diamond design which has a broadening effect on the figure. Plan which crosswise stripe you want at the widest part of the hip, too.

Match plaids at the seven-inch hipline.

Cut with the grain.

Cut the waistband exactly on grain. The waistband in plaid fabrics may be cut either on the lengthwise or on the crosswise grain, depending on the design.

Cut the lining after the first fitting in case the pattern has to be altered.

7. Marking

Use tailor tacks[15] to mark all darts, tucks, crotch seams, and outseams at waistline and hem. Mark original seamline on crotch seam with hand basting.

Stay stitch the waistline 1/8 inch outside the seamline. Stay stitch the seamlines at the placket, the length of the zipper. Do not stay stitch the crotch seam.

Baste[16] the lengthwise and crosswise grainlines with contrasting colored thread if the fabric is not a plaid. Put the crosswise grainlines at a right angle to the lengthwise grainline at the seven-inch hipline marking, the crotch, and the knee.

Press the creases before basting the garment together. Fold on the lengthwise grainline and baste. When pressing the crease, lay the fold on a plaid press cloth and use a line in the fabric as a guide to help you keep the crease straight. If you do not have a plaid press cloth, lay the fold against a yard stick and press. In the front, press only to the point of the dart. In the back, press up to the crotch level as in Figure10.7.

8. Fitting

Read the following suggestions and those in the pattern guide sheet before fitting the garment.

Check the grainlines, ease, balance, and the position of all seams when fitting the garment. Remember, drag[14] lines point to the source of the trouble. Start at the seven-inch grainline first and fit the front, then the back. The crosswise grainline at the seven-inch hipline must be kept parallel to the floor. The creases in each leg must be perpendicular to the floor. Pants have four centers, whereas skirts only have two.

专业词汇

15. tailor tack 线钉

16. baste vt. 粗缝

6. 纸样排版和裁剪

参考排版说明表。一个方向设计和绒毛织物裁剪时纸样在同一方向放置。如果使用格子面料，要决定哪一条纵向线条在侧缝。有时在腰部和腿部做必要的劈势，引起线条成菱形图案，在人体上产生扩张的效果。也要计划你想要的哪一条横向条纹在臀部最丰满的位置。

前后片在7英寸的臀围线处格子面料要对齐。

沿着纱线裁剪。

沿着丝缕精确地剪裁裤腰。裤腰依据设计决定在格子面料的竖向或横向上裁剪。

在第一次试穿后裁剪腰里，以防纸样有改动。

7. 标记

使用钉线标记所有省道、折裥、裆缝线和在腰围线和底边线处的侧缝线。用手缝线标记最初的裆缝线。

腰围线侧缝处的1/8英寸、口袋处的缝线和拉链长度不缝合。缝合裆缝。

如果不是格子面料，用对比色线粗缝经丝缕和纬丝缕。纬向引导线在7英寸臀围线、上裆线和膝围线处垂直于纵向引导线。

在粗缝衣片之前先熨烫前后挺缝线。折叠挺缝线并粗缝。当熨烫挺缝线时，将折叠的衣片放在一块格子面料上熨烫，使用格子面料作为引导，帮助你保持挺缝线笔直。如果你没有格子熨烫布，就将折缝依靠着一码的直尺熨烫。在前片上只熨烫到省尖。在后裤片上只熨烫到上裆线上（图10.7）。

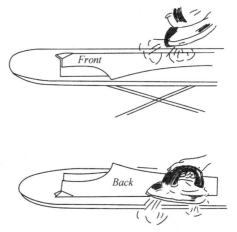

Figure 10.7　Press the front and back grainline
图 10.7　熨烫前后挺缝线

8. 试穿

在试穿服装之前，阅读下面和纸样指南表中的建议。

在试穿的时候，检查挺缝线、松量、平衡和所有缝的位置。记住，牵扯线条指向产生麻烦的源头。首先从7英寸臀围线开始检查前身的合体度，再检查后身的合体度。在7英寸臀围线那里的横向丝缕线保持与地面平行。每条腿上的挺缝线要垂直于地面。裤子有四条中心线，而裙子只有两条。

通用词汇

⑫ layout *n.* 布局　　　　　　⑬ napped *adj.* 有绒毛的　　　　　⑭ drag *vt.* 拖拽

Adjust the darts before fitting the outseams. The darts may be shifted to the area where they are needed. Mini-darts, about two inches long, may be used between the large darts and the side seams.

A slight garment bias in the center back seam is acceptable. Sometimes it is impossible to match plaids at the inseams. Just remember these are the least noticeable seams in the garment. Pants may be tapered in the fabric. Indicate the amount tapered on the pattern.

Chart 10.2 Common fitting problems

Problem	Correction
Crotch too long	Fit out excess length with an even tuck pinned all the way around approximately at the seven-inch hipline marking. Make alteration of this amount in the pattern. Recut the garment using the altered pattern.
Crotch too short	Adjust the waistline seam by setting the lower edge of the fitting band above the stay stitching on the waistline.
Crotch too tight	Let out inseam at point of crotch. Taper new seam down to original inseam seamline.
Wrinkles[15] radiate from the crotch in front	Pants may fit too tightly in hip area. Release outseams. Crotch width may be too small. Let out inseams of both back and front.
Crotch too loose	Take in inseam at point of crotch. Taper new seam down to original inseam seamline.
Waistline pulls down at center front or back	Set fitting band higher at center of waistline, tapering to original seamline at sides.
Horizontal wrinkles across back below waistline	Pants may be too tight. Let out the outseams or back crotch seam.
Saggy[16] seat	Caused by sway back or flat seat. Fit out excess fullness with a tuck beginning at the crotch seam, tapering to nothing at the outseams. The crotch seam may need to be taken in at the waistline and tapered to the original seamline about half way down. The inseam of back leg may also be taken in to decrease width. Shorten darts.
Side seams swing forward	Caused by a prominent abdomen[17]. Lower waistline seam at front crotch seams, tapering to nothing at outseams. Raise the upper end of back crotch seam until outseam is perpendicular to the floor. Taper waistline seam to original seamline at sides. Change tapered darts to tucks.
Side seams swing backward	Caused by a large derriere[18]. Raise the upper end of the front crotch seam until outseam is perpendicular to the floor. Taper waistline seam to original seamline at sides. If necessary, let out inseam of back leg at point of crotch.

在试侧缝之前调整省道。省道也许要调整到需要的位置上。大省和侧缝之间也许应用微型省，长度大约2英寸。

在后中心线微微的斜是可以接受的。有时格子面料在下裆缝很难吻合。记住服装上这些最不引人注意的接缝。裤子也许在面料上劈进，在纸样上标记劈进的量。

表 10.2　一般的合体问题

问题	修正
裆太长	在7英寸臀围标记线那里均匀别去相等的量。在纸样上修改掉这些量，使用修改过的纸样重新裁剪服装
裆太短	将试衣带的下端边缘放置在未缝合的腰围线上方，调整腰围线缝
裆太紧	放开裆部顶点处的下裆缝。向下画出新的下裆缝渐渐地与原来的缝连接
前裆呈放射状皱褶	裤子在臀围区域也许太紧，放开侧缝。也许裆缝宽度太小，放出前后片的下裆缝
横裆太松	在裆端点收进下裆缝。向下画出新的下裆缝，渐渐地与原来的缝连接
腰围线在前中心或后中心处向下拽	在腰围线前或后中心上方增加高度，渐渐地与原来的侧缝线斜向连接
后臀腰围线以下横向皱褶	裤子可能太紧。放出侧缝或后裆缝
臀部下垂	后裆缝太斜或太直。在裆缝线上折去多余的量，从裆缝开始折叠量逐渐变小，在侧缝处没有折叠。也许在腰围线那里的裆缝处折叠，也许在裆缝一半的地方折叠。后片腿部下裆缝也可能折叠，减小宽度。缩短省道
侧缝向前偏移	由突出的腹部引起。在前裆缝处降低腰线，向侧缝画斜线。提高后裆缝顶端，直到侧缝与地面垂直。在腰线上画斜线，与原来侧缝腰线重合。将锥形省改为折叠褶
侧缝向后偏移	由大的臀部引起。提高前裆缝顶端，直到侧缝与地面垂直。在腰围线上画斜线与原来腰线在侧缝重合。如果需要，在后裆端点处放出下裆缝

通用词汇

⑮ wrinkle *vi.* 起皱纹　　　⑰ abdomen *n.* 腹部　　　⑱ derriere *n.* 臀部
⑯ saggy *adj.* 下垂的

Continued chart 10.2

Problem	Correction
Pants pull down in back when you sit down	The back crotch may not have a deep enough curve to give sitting room. Re-shape crotch seam by cutting a deeper curve; let out the outseams. Or, cut and spread pattern on hipline. See Figure 10.8.
Crease lines hang askew[19]	Caused by figure irregularities, i.e., high hip, bowlegs[20], etc. Choose fabrics which do not have horizontal lines.

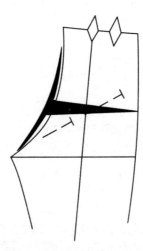

Figure 10.8　Cutting and spreading pattern on hipline
图 10.8　在臀围线处剪切并拉展

表 10.2 续

问题	修正
坐下来时裤子向下拽	后裆缝弧度不够，坐下时没有足够的空间。重新画弧度大的裆部弧线；放出侧缝。或者在臀围线剪开和展开纸样（图 10.8）
挺缝线悬垂不直	由非常规的体型引起，例如臀围偏上、罗圈腿等。选择水平方向没有线条的织物

通用词汇

⑲ askew *adj*. 歪   ⑳ bowlegs *n*. 罗圈腿

Passage 11

Bodice Pattern

There's more than one way to get a perfect pattern for a closely-fitting bodice, such as fitting an armhole (also known as an armscye[1]), including fitting the bust dart, shoulders and side seams. The only way to successfully fit an armhole is to use muslin.

Muslin is a test garment in inexpensive fabric, initially sewn without facings or edge finishes. You pin-fit the muslin right on the body. The object is to make the fabric skim① the body with no signs of wrinkles or strain② lines.

Follow this order when fitting the armhole: bust, back, underarm[2], shoulder seam placement and slope, shoulder point to underarm, and side seams. Then adjust the pattern tissue using the fitted muslin as your guide. Such a completed pattern can be used as a reference to position darts and establish the armhole shape in future patterns. You'll end up with a perfect pattern for a closely-fitting bodice with or without set-in fitted sleeves.

Very different figures people with identical bust and over-bust measurements often fit into the same size, but the style patterns are differently. The shape of your body depends on where you carry your flesh and dictates the shape of an armhole. A muslin is the testing ground—it's the perfect place to sort out fit issues (Figure11.1).

1. Glossary of terms

Use this glossary in Figure11.2 to help navigate your pattern and understand the fitting process. Key landmarks are identified on these pattern pieces. The dotted lines indicate possible fitting sites.

专业词汇

1. armscye *n*. 袖窿　　　　2. underarm *adj*. 腋下的

通用词汇

① skim *v*. 使掠过　　　　② strain *vt*. 拉紧

第 11 课

衣身原型纸样

有好几种方法取得完美的上衣衣身纸样，比如通过合体的袖窿，包括通过合体胸省、肩省和侧缝取得。要想成功地获得合体袖窿，唯一的方法是使用坯布。

坯布是一种便宜的试验服装的面料，最初缝合时不用贴边或拷边。你在身体上用大头针别合坯布正面，其目的是使面料在身体上服帖而没有皱褶或拉拽线条。

在操作合体袖窿时遵循这样的顺序：胸、背、腋下、肩缝位置和肩斜、肩端点到腋下和侧缝。然后使用合体的坯布指导你修正纸样。这种修正过的纸样可以指导省的定位，为将来纸样确定了袖窿形状。最终你将得到一个完美的衣身纸样，它有或没有袖子，但非常合体。

不同体型的人，她们有相同的胸围和乳上胸围尺寸，适合相同尺寸，但是款式纸样不同。你身体的形状依赖于你哪个部位长肉，也就决定了袖窿的形状。坯布是一种基础试验——是辨别何处存在合身问题的好方法（图 11.1）。

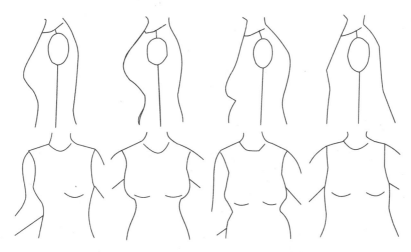

Figure 11.1　Same measurements, very different figures
图 11.1　相同尺寸、非常不同的体型

1. 术语

使用图 11.2 中的术语可以帮助你制作纸样，理解试衣过程。纸样上标记了一些关键点，虚线表示可能做修改的部位。

2. Preparation

You'll need a person, a pattern, and some woven fabric. Select a fitted blouse[3] pattern intended for woven fabrics that includes bust darts in the side seam or armscye. Make sure the finished bust measurement printed on the pattern is between 1 ½ inches and 2 ½ inches greater than your actual full bust measurement to allow enough wearing ease. (Wearing ease is the difference between your body's measurements and the finished garment's measurements, which is necessary so you can move in the garment.) Use a stable woven fabric without spandex[4], and follow the pattern to make your sleeveless, collarless muslin.

Sew the seams using a long machine stitch, and use a thread color that contrasts with the muslin fabric so that you can easily see to clip and release seams during fitting. You'll need an assistant for the fitting process. Plan on making several muslins to get the right fit; the results are well worth the effort.

3. Using muslin

Muslin tells you more than your measurements alone do. The first muslin serves as the rough draft for blocking out the major fitting changes. If you're a beginner, it's better to make more muslins with fewer changes to each than to try too many changes at once.

Don't try fitting muslin that's a size too small because the tightness distorts the overall fit. For example, if you get a strain line between the bust points (Figure 11.3), start over using a larger size pattern.

Figure 11.3 a too-small pattern cause a strain between bust points
图 11.3 太小的样板在两乳之间产生紧绷

专业词汇

3. blouse *n.* 女衬衫   4. spandex *n.* 斯潘德克斯弹性纤维

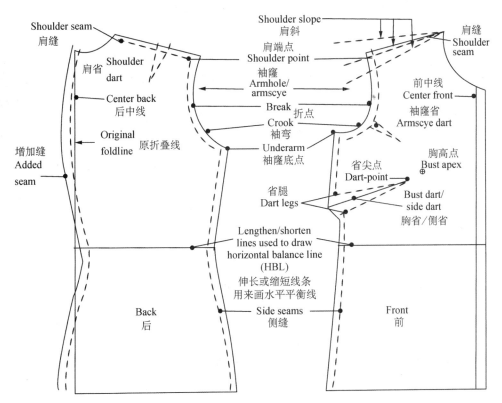

Figure 11.2 Glossary of terms
图 11.2 术语

2. 准备

你需要一个人、一个纸样和一些梭织面料。为了制作梭织面料衬衫，选择一个包括侧胸省或袖隆省的合身纸样。确保印在纸样上胸围的成品尺寸比你实际胸围尺寸大 1 ½ 英寸到 2 ½ 英寸，目的是允许有足够的松量（松量是身体尺寸和成品服装尺寸之间的差，使你穿上这件服装后可以作必要的活动）。按照这个纸样，使用稳定没有弹性的梭织面料，制作无袖、无领的坯布衣身。

用长针脚缝合，并用与坯布成对比色的线，便于你在试衣的时候能清楚地看到修剪和放开的缝份。在合体过程中你需要一个帮手。为了取得很好的合体性，要计划做几个坯布纸样，从而取得满意的结果。

3. 使用坯布

坯布能告诉你比测量还要多的数据。第一块坯布用作打草样，大体画出主要改变的地方。如果你是一个初学者，最好准备几块坯布，每一块上作较少的改变，而不是一次做很多改变。

合体用的坯布尺寸不要太小，因为太紧了会使整体变形。例如，如果在两个乳点之间有拉紧的线条，重新使用一个大尺寸的纸样（图 11.3）。

It's important to make sure the bodice stays level around the body and doesn't dip in the front or back during the fitting. You can draw a horizontal balance line (HBL), on the face of the muslin so you have a point of reference that you can easily see while fitting (Figure 11.4). You can use the lengthen/shorten line between the waist and underarm rather than at the hem as your HBL. While fitting, periodically check the level on the front and back HBL. If the back HBL dips, pin a wedge out of the upper back to level the line.

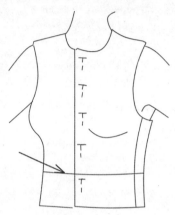

Figure 11.4   Use a horizontal balance line (HBL) to help you keep your muslin level
图 11.4   使用水平平衡线（HBL）有助于保持坯布水平

4. Fitting assessment

Stand back and read the muslin to assess the fit. To get started, study the general fit of the muslin and make any obvious adjustments. If the side seams strains over the hip, open both side seams from the hem upward until the muslin falls nicely (Figure 11.5). If the shoulder seam is too loose, pin out the excess fabric. The muslin should fit without strain but not be loose and baggy. Remember to get a pattern that reflects the shape of the body. You'll want to develop a fitting muslin that fits like a second skin—snug but not tight. Add design and additional wearing ease later.

It's usual to fit only one side of the muslin after making the HBL level. If the person is particularly asymmetrical, overfitting can accentuate an uneven body. In general, if one side of the bust is larger, fit the larger side; if one shoulder is higher, fit the higher side, and adjust the low shoulder with a pad.

5. Fitting the front

Assess how the muslin fits at the bust. Darts contour the fabric to accommodate the swell[③] of the bust while keeping the garment looking trim. Anyone with a full A-cup or larger benefits from properly placed bust darts, which make the center front of the garment fall straight to the waist and not swing away from the body. This results in a more flattering silhouette. The correct bust dart also keeps the armhole from gaping.

合体时，确保衣身在身体上保持水平，不要前面或后面下沉。你可以在坯布的表面上画一根水平方向平衡线（HBL），这样你有一个参考点，在合体的时候很容易看清（图 11.4）。你可以在腰部和腋下之间用一根伸缩线，而不是将底边作为 HBL 线。要经常检查前后 HBL 线呈水平状态。如果后面的 HBL 下沉，在后背上方折叠楔形，直到呈水平状态。

4. 合体性评估

挺直身体，研究坯布，评估合体性。开始时，研究坯布的大致合体程度，对那些明显的地方作修正。如果侧缝在臀部拉紧，从底边向上剪开侧缝，直到坯布下垂呈良好状态（图 11.5）。如果肩缝太松，别去多余的面料。坯布必须合体，没有拉扯，也不松弛和下垂。记住获得一个能体现身体形状的纸样。你要开发一个合身体的坯布衣身，就像第二层皮肤那样——紧身但不勒身。然后再添加设计和额外的穿着放松量。

在标记了 HBL 水平线以后，通常只在坯布的一侧试验合体性。如果一个人体型特别不对称，过于合体反而显示出不均衡的身体。通常，如果一侧的胸部大，就合体大的一侧；如果一侧肩部偏高，就合体高的一侧，低的一侧用衬垫修正。

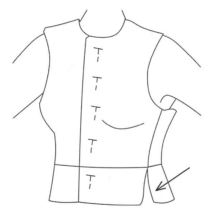

Figure 11.5　If your side seams strain, open them over the problem area
图 11.5　如果侧缝拉紧，应放开直到平服

5. 前衣片合体性

评估坯布在胸部的合体度。省道缝合呈弧形，适合胸部隆起，保持服装圆润。有着丰满 A 罩杯或大胸部的人，需要放置恰当的胸省，使服装前中心直到腰部都能保持垂直状态，而不是弯在身体上，这样会产生更加好看的廓型。正确的胸省使袖窿与身体之间没有空隙。

通用词汇

③ swell *n.* 凸起的形状

Some women have bust tissue under their arms; Other women carry it only in front. Experiment with positioning the angle of the dart until it's most flattering. The dart placement alone can visually slenderize the figure.

You can use side darts in a base pattern pinching[4] out a dart in the armscye to eliminate the gape and then moving the dart to a better location later is a good approach. Read the muslin, and use your fingers to ease up any drag line—in this case, a diagonal fold of fabric occurs between the bust apex and the side seam—into the side bust dart (Figure11.6).

The way you drape a bust dart influences how it flatters the body. Use the dart as a design element that directs the eye to your advantage. Set the point of a dart closer to the apex for a smaller bust, farther away for a full bust. Don't be surprised if your dart seems larger than usual, as long as you're getting a smooth fit. A very large dart may be needed to fit a very full bust. If this causes an unattractive bubble at the dart point, two parallel darts will solve the problem. Pin in the required dart(s) that best fit the bust (Figure11.7).

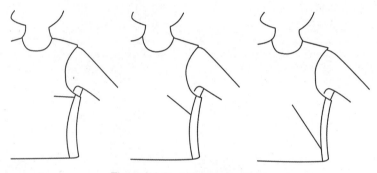

Figure 11.7 side-seam darts
图 11.7 侧缝省

Increasing the size of a side-seam dart lowers the front armhole. Fill the vacancy[5] with a small piece of fabric to bring it up to the original height and redraw the armhole seamline (Figure11.8).

Stand back and evaluate whether the dart point hits in a pleasing place on the bust. If the dart is too low, it looks matronly[6], and if the dart is too high, it can look unflattering as well. The way you drape the dart influences how it flatters the body. Decide whether a slanting or straight dart fits you best.

6. Fitting the back

Dart corrections on the back shouldn't end in the armhole. After you've adjusted the bodice front, check the back for excess fabric or undue[7] strain at any point and assess the fit in the same way you did on the front. If the back armhole gapes, release the side seam and push the side back toward the front to diminish the gape and redraw the side seam (Figure 11.9).

If there is considerable roundness in the back, adding darts at the shoulder seams or even adding a center-back seam for extra curvature is a good solution. If an armscye dart is needed to

一些女性的胸部组织在手臂下方；还有一些女性在前身。反复定位省道的角度，直到最漂亮为止。仅靠省道的定位就能从视觉上产生苗条的体型。

在基础纸样上，你可以使用侧省，在袖窿上别出一个省，消除空隙，然后将这个省道移到更好的位置上，这是一种好的方法。审视坯布衣身，使用你的手指移动任何拖拽的线条——在这种情形下，在胸部最高点和侧缝之间产生的对角线方向的折叠——形成侧胸省（图11.6）。

捏胸省的方法影响省美化身体的程度。将省作为一种设计元素，将视线引导到你的优势之处。对于较小的胸部，省顶点离胸高点近一些，较大的胸部则较远些。如果省看上去比正常的大，不要惊讶，只要获得圆润合体就行。特别丰满的胸部适合特别大的省。如果在省尖点产生不好看的凸起，就用两个平行的省解决这个问题。在胸部最合适的部位别出需要的省（图11.7）。

增加侧缝省的尺寸，前袖窿将会降低，用一块小的面料填补到原来的高度，重新画袖窿弧线（图11.8）。

挺直身体，评估省尖是否在胸部合适的位置上。如果省太低，看似老大妈，如果省太高，看上去又不好看。别合省道的方法影响穿在身体上的美感。决定斜向还是直向省最适合你。

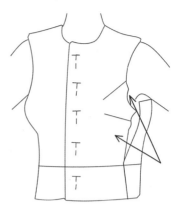

Figure 11.6　Ease excess fabric to form darts
图11.6　多余的面料形成省道

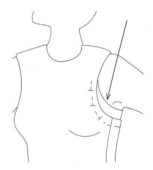

Figure 11.8　Raise the depth of the armhole
图11.8　抬高袖窿深

6. 后衣片合体性

后衣片袖窿上不应该有省道。在调整了前衣片后，应该检查后衣片上冗余的面料或在某一点过度的拉扯，用评估前衣片的方式评估后衣片。如果后袖窿有空隙，放开侧缝，将后衣片向前推移，去除空隙，并重新画侧缝线（图11.9）。

如果在后片上有相当大的鼓泡，在肩线上添加省道，甚至增加后中缝，即增加弯曲程度，这是一种好的方法。如果需要在后袖窿上设置省，以适应后背肌肉的需求，就用公主线

通用词汇

④ pinch v. 捏
⑤ vacancy n. 空缺
⑥ matronly adj. 主妇似的
⑦ undue adj. 过度的

fit a muscular back, incorporate the dart amount in a princess line (back armscye darts are not traditionally used), as shown at Figure11.10. The object is always to reduce any excess fabric in the circumference[5] of the armscye.

Figure 11.9 If the back armhole gapes, reposition the side seam

Figure 11.10 Add princess seams and a center-back seam to accommodate extra back curvature

图 11.9 如果后袖窿有空隙，重新画侧缝线

图 11.10 添加公主线和后中缝线，适应后背的曲面

7. Fitting the armhole

Most patterns are cut too low under the arm. Now that you've draped darts to match your curves and removed excess fabric from the back of your muslin, the shape of your armhole has probably changed. Deciding where the armhole hits under the arm is partially personal preference. Remember that an armhole cut high up under the arm is generally more comfortable because it allows a greater range of movement in a garment with sleeves; this is often counter-intuitive[8] to a beginning fitter. A sleeveless garment is only ½ inch higher under the arm than a fitted garment with a sleeve.

If the armhole is cut too low under the arm, add a piece of fabric, and draw in a new depth. Alternatively, make a note to raise the underarm a specified amount on your pattern tissue. If you hold a ruler under your arm as high as is comfortably possible, the underarm seamline should fall barely below where the ruler is touching the flesh.

The shoulder seam should lie along the top of the shoulder at a place that balances the body front to back and follows the natural slope of the shoulder (Figure11.11). The shoulder point falls on the shoulder seam at the exact place the arm and shoulder come together—at the dent[9] that forms when you lift your arm.

On the muslin, draw a line that falls from the shoulder point to the "crook"[10] of the arm (where the arm attaches to the body) and then runs under the arm at the "break of the arm" (where the curve begins to go under the arm). Draw the curvature of the armhole on the muslin to follow the body curvature from the shoulder around the arm on the front and the back.

处理省量（一贯以来不使用后袖窿省），如图 11.10，目的就是在袖窿周围减少任何冗余的面料。

7. 袖窿合体性

很多纸样在手臂下方裁剪太低。既然你已经设定了省道适应你的身体曲线，并在坯布的后衣片上去除了冗余的面料，袖窿的形状可能已经发生变化。袖窿在手臂下方的什么位置上，部分地由个人喜好决定。记住，在手臂下方袖窿高一点更加舒适，因为装了袖子以后，活动空间更大，这与初学者的预料是相反的。无袖服装比合体有袖服装的袖窿在手臂下方高出 ½ 英寸。

如果手臂下方的袖窿太低，补一块面料，重新画袖窿深。或者，在你纸样上标注腋下提高的量。如果在腋窝下方放一把尺，尽量高，但又舒适，那么腋下袖窿弧线略微低于尺触及肌肤的位置。

肩线应该在肩顶部，前身和后身平衡的位置上，并且吻合肩部的自然倾斜度（图 11.11）。肩端点在肩线上，准确的位置在手臂和肩的交接点——举起手臂时产生的凹痕处。

在坯布上，从肩端点开始向下到臂弯处（手臂触及身躯的地方）画线，然后通过手臂下方，到手臂断点（手臂下方弧线开始的地方）。在坯布上，依据身体弧线围绕手臂在前身和后身画袖窿弧线（图 11.12）。

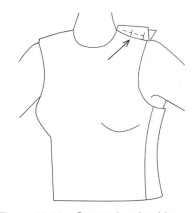

Figure 11.11 Center the shoulder seam and correct the slope

图 11.11 肩线放置中间位置和纠正肩斜

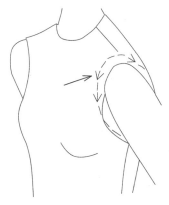

Figure 11.12 From the shoulder point, follow the body curvature and draw a new armhole

图 11.12 从肩端点，按照身体曲线画新的袖窿线

专业词汇

5. circumference *n.* 周围

通用词汇

⑧ counter-intuitive 与正常预期相反的 ⑨ dent *n.* 凹痕 ⑩ crook *n.* 弯曲

Check that the side seam hangs straight. Make adjustments by releasing and repinning the seam so that it's perpendicular to the floor. Assess if it divides the side of the body attractively.

At last, transfer the muslin alterations to the pattern. Mark the seam and dart lines directly on the muslin, and follow any instructions noted on the muslin during the fitting. Use a permanent marker, and always mark and concentrate on the actual seamlines. To reduce confusion, ignore seam allowances until later. The muslin is now a road map of the changes needed on the pattern.

检查侧缝垂直服帖。通过放开或重新别合作调整，使得侧缝垂直于地板。评估侧缝是否在前后身理想的位置。

　　最后，转移修改的坯布到纸样上。在坯布上直接标记缝线和省道线，并将坯布所标记的说明转移到纸样上。使用不褪色记号笔，反复做标记，专心于实际缝线。为了避免混淆，暂时不考虑缝份，到后来再考虑。现在坯布成为纸样上的修改路线图。

Passage 12

Sleeve Pattern

Exceptional[1] garments often are easy to identify by examining the quality of the sleeves. There are many different methods to draft sleeve pattern. Here four basic techniques of sleeve are introduced.

1. Enlarging the sleeve and raising the sleeve cap

Start by measuring your arm circumference at your mid upper arm and add 1 inch for ease for a tee-shirt sleeve or 1½ inches for a cardigan[1]. Then measure the width of the sleeve pattern less seam allowances at the CD line as shown on Figure12.1, which is 6½ inches down from the sleeve cap. Compare your arm measurement with ease to the sleeve measurement. If the difference is 3½ inches or less, you can modify your sleeve pattern with these instructions. If the difference is greater than 3½ inches, do not use this technique only, you may have to add some to the sleeve seam as well.

Draw a copy of the sleeve cap on the colored tissue (Figure12.1). Be sure to mark back, front and shoulder marks. Set the colored tissue sleeve cap drawing aside for future use. Do not attach to the master sleeve pattern yet. On the Master Pattern, mark the following points as shown on Figure12.1.
- Point A is the top of the sleeve cap, which is beside the shoulder mark.
- Draw a vertical line from point A that is perpendicular to the hemline and mark Point B.
- Points C and D are 6½ inches down from the sleeve cap at the side seams.

Draw a line from C to D, which is perpendicular to the AB line. Cut the Master Pattern from B up to A – to but not through – the stitching line, then down from A to stitching line (to but not through) to create a paper hinge (Figure12.2). Cut from AB line to C and AB line to D – to but not through – the stitching line. Then cut from the edges at C and D to but not through the stitching line to create a paper hinge.

Place a 5-inch by sleeve length strip of gridded[2] pattern paper underneath the AB line of the Master Pattern. Lay one side of the AB line at CD (cut edge) on a vertical grid line that is ½ inch from the edge of the gridded pattern paper, so that the hemline is on a horizontal line. Tape the vertical line

---

专业词汇

1. cardigan *n.* 开襟羊毛衫

第 12 课

袖子纸样

合体服装可以轻易地通过检验袖子的质量来判断。有很多画袖子纸样的方法，下面介绍四种基本的袖子纸样技术。

1. 放大袖子和提升袖山

首先在你的上臂中部测量臂围，T恤袖子增加1英寸松量，对衬衫袖子增加1½英寸。然后在袖纸样上测量CD的宽度，并减去缝份，CD线在袖山下方6½英寸（图12.1）。比较你的包含松量的手臂尺寸和袖纸样的尺寸。如果差小于等于3½英寸，你可以根据这些讲解修改你的袖子纸样。如果差大于3½英寸，除了采用这种技术，你还需要在袖缝上增加一些量。

在有色棉纸上拷贝袖山弧线（图12.1）。在袖纸样的后、前和肩做标记。将画好袖山的有色棉纸放在一旁等待使用。现在还不要粘到袖子原型纸样上。在袖子原型纸样上按照图12.1显示做标记。

- A点是袖山的顶点，在肩点旁。
- 从A点画一条垂直线，它垂直于底边，并且标记B点。
- C点和D点在袖侧缝上，从袖山顶点向下6½英寸。

从C到D点画一条线，垂直于AB线。在原型纸样上，从B向上剪开到A，但不要剪断缝线，这样，从A到缝线纸样可以转动（图12.2）。从AB线剪到C，从AB线剪到D，但是不要剪断缝线。然后在C和D的边缘剪开，但是不要剪断缝线，产生纸样的转动。

将5英寸宽、袖子长度的长条网格纸，放在原型纸样的AB线下方。将剪开的到CD线（剪切边缘）的AB线的一边放在距离网络纸边缘½英寸的垂直线上，使得袖底边线在水平线上。从袖子底边（B点）沿AB线垂直向上用胶纸带将纸样和网格纸粘上，CD线下方1½英

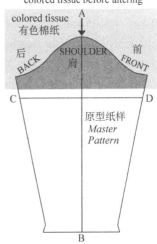

Figure 12.1　The sleeve pattern
图 12.1　袖纸样

通用词汇

① Exceptional *adj.* 优越的　　　　② grid *n.* 格子

to the gridded pattern paper from the hemline (Point B) up to 1½ inches from the CD line. Do not tape all the way to the top yet, just from CD line down.

Increase the opening of the AB line by the amount that you need to add for the sleeve width, which could be up to 3½ inches. The gridded pattern paper makes it easy to accurately add the same amount so that the cut lines are parallel and the hemlines on both sides are on the same horizontal line. Tape this second vertical line to the gridded pattern paper, from the hemline (Point B) up to 1½ inches from the CD line.

Notice what this does to the Master Sleeve Pattern above the CD line. One cut edge of the CD line will overlap with the other cut edge of the CD line. Be sure that the paper hinges on the CD line allow the Master Pattern to lay flat. Tape down the remainder of the AB line and the overlapped CD lines. It should now look like Figure 12.3.

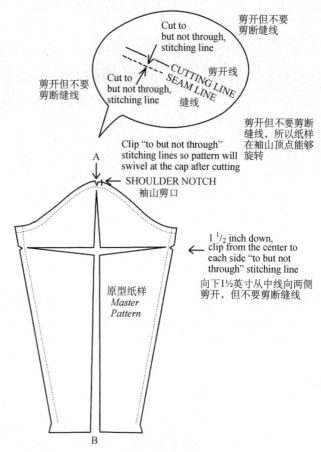

Figure 12.2 Clipping the master pattern
图 12.2 剪开原型纸样

Now get the colored tissue where you traced a copy of the sleeve cap. Lay the tissue paper behind the Master Sleeve Pattern so the fronts and backs are together. Align the shoulder marks on the tissue and the Master Sleeve Pattern, so the tissue shoulder mark is directly above the shoulder mark on the Master Sleeve Pattern. It is possible that the tissue drawing of the sleeve cap may not match the enlarged Master Sleeve Pattern at the side seam. Sometimes, the tissue will be ½" narrower than the Master Pattern on each side. This is not a problem — use the width of the enlarged Master Sleeve Pattern. Tape the colored tissue to the Master Pattern. Cut around the sleeve cap on the colored tissue (Figure 12.4).

Your Master Sleeve Pattern has now been enlarged for a larger upper arm. If your Master Pattern was slightly wider than the colored tissue, ease that slight amount into the armhole.

If your arm measurement with ease is smaller than the Master Sleeve Pattern, you can follow the same process. However, instead of adding the gridded pattern paper underneath, you will overlap the two sides of the AB line for the amount that you want to narrow your Master Sleeve Pattern.

寸结束。此时还不要粘贴到顶端，只是CD线下方。

扩大 AB 线的展开量，直到你想要的袖子宽度，可以扩大达 3½ 英寸。网格纸能轻易地、准确地增加相等的量，使剪开的 AB 线平行放置，剪开的底边线放置在相同的水平线上。将第二根剪开的 AB 用胶带粘贴到网格纸上，从底边（B点）向上到 CD 线下方 1½ 英寸处。

留心这样做以后，在袖子原型纸样 CD 线上方出现的情况。CD 线的一条剪切边与另一条剪切边重叠起来。转动 CD 线，将原型纸样摊平。用胶带粘贴 AB 线的其余部分和 CD 线的重叠部分。现在应该像图 12.3 那样。

现在取回你拷贝袖山弧线的有色棉纸。将棉纸放在袖子原型纸样下方，使袖子的前后彼此重叠在一起。对齐棉纸和袖子原型纸样上的肩点，使棉纸的肩点在袖子原型纸样的正上方。也许棉纸上画的袖山弧线与扩大了的袖子原型纸样在侧缝那里不吻合。有时棉纸的每一边比袖子原型纸样要窄 ½ 英寸。这不是问题——使用扩大了的袖子原型纸样的宽度。用胶带将棉纸粘到原型纸样上。沿着袖山修剪有色棉纸（图 12.4）。

现在你的原型纸样的上臂被扩大了。如果你的原型纸样略微比色棉纸宽些，就将那些量纳入袖窿里。

如果加入松量后你手臂的尺寸小于袖子原型纸样，采用相同的方法。但是，AB 线不是在下方的网格纸增加一定的量，而是要重叠 AB 线的两边，其量就是你想要使袖子原型纸样变窄的量。

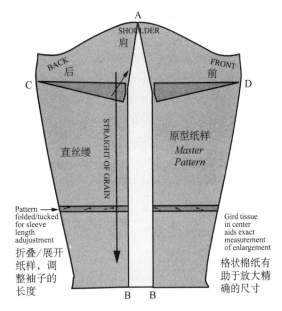

Figure 12.3　opening the AB line
图 12.3　展开 AB 线

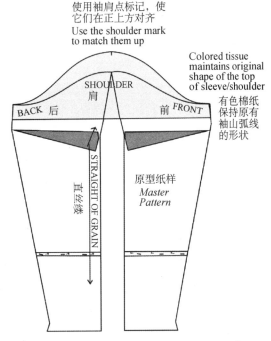

Figure 12.4　Tape the colored tissue to the Master Pattern
图 12.4　将有色棉纸粘贴到袖子原型纸样上

If the master pattern you used is for a knit pattern and has a flatter sleeve cap (as most knit patterns do), you may wish to raise the cap to allow a little more drop to your sleeve. Each person is different as to what they need. For example, measure up the 1½ inches (or your figure) at the shoulder line, mark, and then taper down to finally blend into each side of the sleeve cap (Figure12.5).

One other thing about knit sleeve patterns, a lot of the designers mark the shoulder line at an equal distance from the sleeve side seams. The better drawn patterns have the back sleeve cap longer than the front sleeve cap, as most shoulder seam should be more forward, which means a longer back sleeve.

2. Drafting a cap sleeve

A cap sleeve is more of a decorative detail than a true sleeve ——but it's a great-looking addition to what would otherwise be a simple arm-baring top.

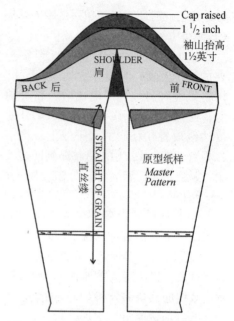

Figure 12.5　The different caps of sleeve
图 12.5　不同高度的袖山

Like its name implies, this sleeve consists of little more than the cap itself, and generally doesn't go all the way around the armhole. As a shoulder detail, it adds a bit of interest and coverage when you want something just a bit more than sleeveless, but less than a typical short sleeve. And it's easy to draft and to sew!

You'll start with your blouse sleeve block (the sketch shows only the upper part of the block, since we're making a very short item). Start by shortening the cap itself by about 1/2". This reduces (or almost eliminates) the ease at the cap; it's not needed in this case (Figure 12.7).

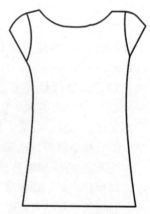

Then mark a point on the grainline about 1½" to 2" above the bicep line. From there, draw a gentle curve to each side of the cap, below the notches (the ends will be about ¾" above the bicep line). This is your hemline (Figure 12.8).

Figure 12.6　A cap sleeve
图 12.6　帽袖

The end result is a very short sleeve, which does not surround the arm. Be sure to mark where the sleeve ends on the front and back blouse pieces!

For a very casual top, you could turn and stitch the hem, but this sleeve looks best with a hem facing. The facing should be applied before setting the sleeve into the armhole.

如果你将原型纸样用于针织面料服装，它的袖山弧线平坦一些（就像大多数针织面料纸样那样），你可能希望抬高袖山使袖子直立一些。不同的人有不同的需要。例如，在肩线那里（或在你身体上）向上量 1½ 英寸，做标记，最后画顺袖山的每一边（图 12.5）。

针织面料袖子纸样上还要考虑的是，很多设计师画的袖山线与袖两边侧缝距离相等。而比较好的袖子纸样应该是后袖山长于前袖山，因为大多数肩缝应该是前倾的，也就意味后袖山更长一些。

2. 画帽袖纸样

帽袖比真正的袖子更具有装饰细节——它是外观漂亮的附加物，否则就是一件简单的上臂裸露的上衣（图 12.6）。

正如它的名字包含的意思，这种袖子就是由一点点袖山构成，通常不是覆盖整个袖窿。作为肩部的一个细节，它增添了一些趣味，比无袖覆盖了你想遮盖的东西，但比正常短袖覆盖面小。并且它的纸样容易画和也容易缝制。

你首先用你衬衫的袖子原型（因为我们画十分短的袖子，图中只显示了原型的上端）。首先将袖山缩减约 ½ 英寸。这将缩小了（或几乎取消了）袖山的松量；在这种情形下不需要松量（图 12.7）。

然后在袖山深线上方 1½ ～ 2 英寸的丝缕线上标记一点。从这点画一条圆顺的弧线触及刀眼（约在袖山线上方 ½ 英寸），并延续至袖山两侧，这条弧线就是帽袖的的底边线（图 12.8）。

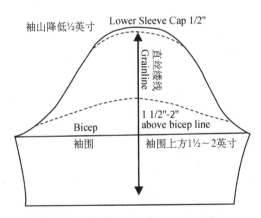

Figure 12.7　Lower sleeve cap ½"
图 12.7　降低袖山 ½ 英寸

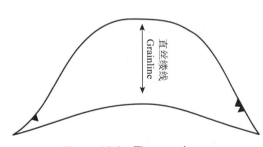

Figure 12.8　The cap sleeve
图 12.8　帽袖纸样

结果就是一个十分短的袖子，它没有围绕整个手臂，但是一定要在衬衫纸样的前后袖窿上标注帽袖的末端点。

如果是十分休闲的上衣，你可以卷边，然后缝线，但是这种袖子还是采用贴边缝制比较好。贴边应该在将袖子安装到袖窿之前先缝制好。

3. Drafting a flutter[3] Sleeve

What is a "flutter" sleeve? It's a full, flowy - and usually short - sleeve that's perfect for a summer top or dress (Figure 12.9).

This is a very feminine look, very airy and cool in warm weather. It can be made long or short, but it's the most "fluttery" when worn short. It's also a variation on a full sleeve - but the fullness is loose and fluid, not gathered into a cuff.

Start with your blouse sleeve block, shortened to about 2-3" below the bicep. Draw vertical lines dividing the sleeve into several sections, about 1½" wide (except for each end - leave those sections about 3" wide) (Figure 12.10).

Cut along these lines from the hem almost to the sleeve cap. It's best to keep the sections connected at the sleeve cap so they don't get out of order, but it's not a problem is you accidentally cut through the edge - just don't re-arrange the pieces!

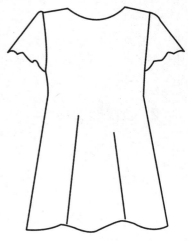

Figure 12.9   The flutter sleeve
图 12.9   飘袖

Spread the pieces apart to create width and fullness at the bottom. For a relatively modest flutter and flare, separate the pieces about 1½" each (Figure 12.11). You can adjust the pieces to add as much or as little flare as you want - for a fuller flare and more drape in the sleeve, spread the pieces about 2" each, until the side seams are almost horizontal. The final shape may seem a bit strange, but it will produce a nicely draped flare at the hem.

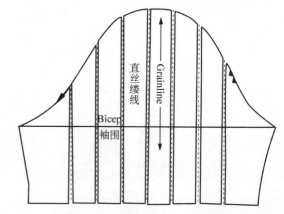

Figure 12.10   Dividing the sleeve into several sections
图 12.10   将袖子划分为几个部分

All of the fullness in this sleeve is at the bottom - there is no gathering or puffiness at the sleeve cap. Give this one a try on your next garden party dress or warm weather top (Figure 12.12)!

4. Drafting raglan sleeve

A raglan sleeve is a sleeve that extends in one piece fully to the collar, leaving a diagonal seam from underarm to collarbone. How to transform a basic sleeve into a raglan one?

3. 画飘袖纸样

什么是飘袖？它是一种宽大、飘逸的——通常是短袖——这种袖子很适合夏天短袖上装或连衣裙（图 12.9）。

它是十分女性化的样式，在炎热的天气里非常透气和凉爽。它可以长些或短些，但短袖最为飘逸。它的宽大程度也可以变化——但这种宽大是松散和飘逸的，而不是用克夫收缩起来。

首先使用你的衬衫袖子原型，在袖山线下方约 2~3 英寸剪短袖子。画垂直线条将袖子分为几个部分，每个部分大约 1½ 英寸宽（两边约留 3 英寸宽）（图 12.10）。

沿这些线条从底边到接近袖山的位置剪开。最好每一部分在袖山那里连接在一起，使它们保持秩序，不过如果你偶尔剪断了也没有问题，只要这些衣片没有放乱。

展开每一片，使底边变宽大。为了比较端庄和飘逸，每一片展开量约为 1½ 英寸（图 12.11）。你可以调节，每一片的展开量可根据自己需求可多可少。要想袖子有比较多的悬垂褶，展开量可以是 2 英寸，袖侧缝线几乎是水平状态。最后的形状可能有点奇怪，但它将在底边产生很漂亮的悬垂褶的。袖子上所有展开量都在袖子底边，在袖山处没有任何抽褶或膨胀。尝试做一件这种袖子的衣服，用于参加下一个花园聚会或炎热天气里穿着（图 12.12）。

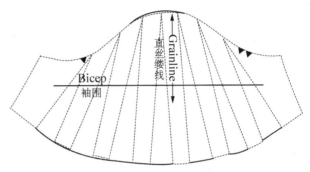

Figure 12.11　Spread the pieces apart to create width and fullness the bottom

图 12.11　展开每一片，使底边变宽大

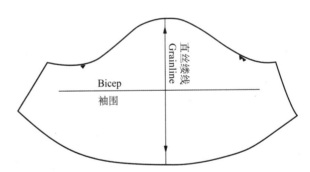

Figure 12.12　The pattern of flutter sleeve

图 12.12　飘袖纸样

4. 画插肩袖纸样

插肩袖是一整片袖片一直延伸到衣领，从腋下到锁骨留下一条对角线。如何将基础袖子转变为插肩袖？

通用词汇

③ flutter *n*. 飘动

1）Temporarily move back shoulder dart. Close bust dart on front bodice, moving it temporarily if necessary (remember this dart can be placed anywhere after the raglan adaption has been made) (Figure 12.13).

2）Trace around the sleeve block and remove all unnecessary ease from sleeve head (Figure 12.14).

3）Lay the front and back bodice onto the traced sleeve matching balance points-leave an approximately ¼" gap between the Shoulder Neck point(SNP)[2] and center of sleeve (Figure 12.15).

4）Draw in your required seam lines, ensuring that they blend into the bodice and sleeve at the balance points (Figure 12.16).

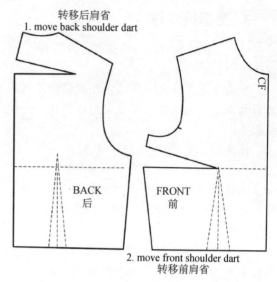

Figure 12.13　Moving front and back shoulder darts
图 12.13　转移前后肩省

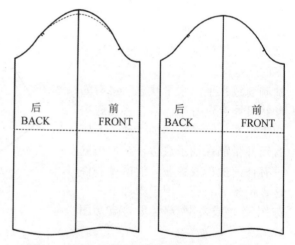

Figure 12.14　remove ease from sleeve head
图 12.14　从袖山去掉松量

专业词汇

2. shoulder neck point (SNP) *n.* 颈肩点

1. 临时转移后肩省。在前衣片上闭合胸省，如果需要的话临时转移（记住在套肩袖画完之后这个省可以转移到任何地方）（图 12.13）。

2. 拓印袖子原型，并且在袖山上去除所有不必要的松量（图 12.14）。

3. 放置前后衣片在拓印的袖子上，与平衡点吻合，在颈肩点（SNP）和袖中心之间留有 ¼ 英寸空隙（图 12.15）。

4. 画你想要的缝线，在衣身和袖子连接的平衡点处要画圆顺（图 12.16）。

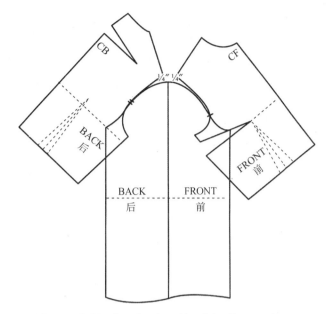

Figure 12.15　Lay front and back bodice on sleeve
图 12.15　将前后衣片放在袖子上方

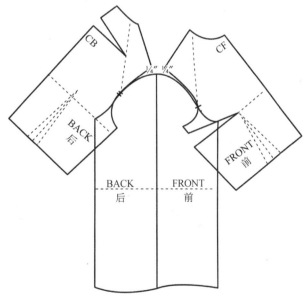

Figure 12.16　Draw in your required seam lines on bodices
图 12.16　在前后衣身上画线

5) Create a dart at the sleeve head of approx. 2" to give a smooth run over the shoulder (Figure 12.17).

6) Trace off new pieces closing temporary back shoulder dart add seam allowance, etc (Figure 12.18).

Note: this sleeve can easily be converted to a two piece raglan sleeve by separating the sleeve on the center line and adding seam allowances.

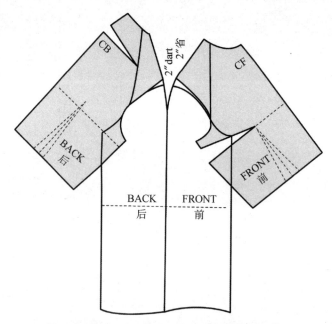

Figure 12.17　Create a dart at the sleeve head
图 12.17　在袖顶点设计一个省

5. 在袖山最高点设计一个大约 2 英寸的省，目的是使肩部圆顺（图 12.17）。
6. 拓印新的衣片纸样，闭合临时的后肩省，增加缝份等等（图 12.18）。
注释：这种袖子可以很容易转变为两片插肩袖，只要沿袖中心线剪开，并且增加缝份。

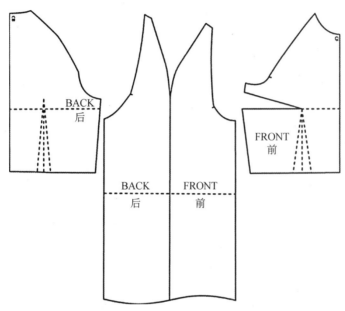

Figure 12.18　Trace off new pieces
图 12.18　拓印新的衣片纸样

Passage 13

Basic Pattern Techniques of Jackets

Basic techniques are the foundation in drafting pattern, which can expand your skills into something amazing.

1. Drafting a notched collar

A notched collar is a signature feature on many blouse, jacket and coat styles for both men and women. This type of collar looks stylish and sophisticated no matter how you sew it!

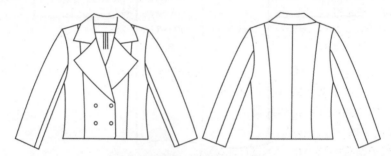

Figure 13.1　Anotched collar
图 13.1　翻驳领

Drafting a notched collar can be a little tricky, but follow these steps and you'll be sewing your own notched-collar garment in no time! We'll start off with some basic collar terminology:

The neckline edge: the edge of the collar that gets sewn to the neckline of the garment.

The collar edge: the outer edge of the collar.

Collar stand: The height of the collar from the neckline edge to the roll line. The height of a collar stand is relative to the neckline. A round, close fitting neckline will have a higher collar stand, while a collar will lay flatter on a wider neck opening.

Roll line: the point at which the collar folds over towards the shoulders.

Top Collar: the upper, outermost layer of the collar.

Undercollar[1]: the layer of the collar that is hidden, laying against the body.

Drafting a notched collar:

Start by taking your personal bodice slopers and measuring the length of both the front and back necklines. Record the measurements (Figure 13.2).

第 13 课

上装基础纸样技术

基础纸样技术可以拓展你的技能，从而达到惊人程度。

1. 画翻驳领

翻驳领不管在男装还是女装中都是很有特色的领子，用于衬衫、夹克和外套。这种类型的领子不管你怎么缝制，看上去时髦和优雅（图 13.1）！

画翻驳领有一点棘手，但是跟随如下步骤，你将能立即缝制你的翻驳领服装。我们首先了解基础领的一些术语：

领口边缘：与服装领围线缝合的领子边缘。

领子边缘：领子的外边缘。

领座：从领口边缘线到翻领线的高度。领座高度和领口线相关。圆形合体的领口线相应的领座较高，而较宽的敞开领口线相应的领座较低。

翻领线：领子向肩部翻折的线条。

领面：领子的上层、最外层。

底领：领子隐藏的那一层，并与身体接触。

画翻驳领：

首先采用你个人的衣身原型，测量前领和后领口线的长度，并记录尺寸（图 13.2）。

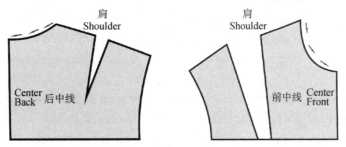

Figure 13.2　Measuring the length of neckline
图 13.2　测量领口弧线长

专业词汇

1. undercollar *n.* 底领

Step 1: Trace out your front bodice sloper onto a large sheet of paper. Draw a line parallel to the center front on your bodice pattern to create a button extension[①]. The width of the button extension can vary. End the line at the point where you want the fold of the collar to end. This point is called the "breakpoint[②]". In Figure 13.3, the jacket's "breakpoint" is just above the top right button.

Step 2: Make a mark 1" out from the front neck-shoulder point. Draw a straight line from the breakpoint, passing through the 1" mark. You can call this line A.

Step 3: Mark a point on the front shoulder measuring ¼" in from the neck-shoulder point. Label point B. Draw a line from this point, extending from the shoulder, running parallel to

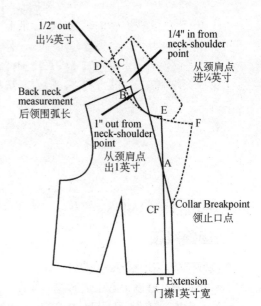

Figure 13.3 Drafting the notched collar
图 13.3 画翻驳领

line A. You can call this line C. (Note: You will also need to remove ¼" from the neck-shoulder point on the back sloper pattern so that the shoulders will still line up when sewn together.)

Step 4: Measure up line C and mark the distance of the back neck measurement. Make another mark ½" from this point, towards the shoulder. Label this point D.

Step 5: Connect point D to point B with a curved line. Re-measure the curved line to ensure the length matches the original back measurement.

Step 6: Line up the straight edge of your ruler with the curve, drawing another line at a precise 90 degree angle from the curve. Draw this line to measure the desired width of your collar. This will be the center back seam of the finished collar.

Step 7: Draw in the collar edge. You can modify this line to create your own collar shape.

Step 8: Measure down the center front line, and mark how low you'd like your neckline to be. You can call this point E. (In this example draft, point E remains at the original center front neckline point.)

Step 9: Draw a line from point B that passes through point E. This line should blend smoothly with the back neck curve, and continue several inches past the center front.

Step 10: Draw a line from the outside collar edge to meet the new neckline. Make a notch on the bodice pattern marking exactly where the collar piece touches the neckline.

Step 11: Draw a curved line from point F to meet the breakpoint.

This draft will now be divided into separate pieces. The left and right front, the upper collar, and the undercollar. The undercollar is cut in two separate pieces, on the bias grain.

步骤1：将前衣身纸样拓印到一张大纸上。在衣身纸样上画一条平行于前中心线的线，那是纽扣的叠门。纽扣叠门的宽度可以变化。你想领子翻折在哪里结束，这条平行线就在那里结束。这个端点也称为"止口点"。在图13.3中，夹克的止口点正好在右衣片第一粒纽扣上方。

步骤2：从颈肩点向外1英寸做标记。从止口点画直线，通过1英寸的这点。你可以称这条线为A。

步骤3：在前肩线标记一点，即从颈肩点沿肩线向内¼英寸，标记B。从这点画一条线，从肩线向外延伸，与线A平行。你可以称它为线C（注意：你还需要在后衣片的肩斜线上向内移动¼英寸，在缝合时，肩线仍然对齐）。

步骤4：在线C上从B点向上量取后领口弧长的尺寸。从这点向肩的方向偏离½英寸，标记D点。

步骤5：用弧线连接D点和B点。重新测量这个弧线，确保长度与原来后领口弧线尺寸相等。

步骤6：用带有弧线的尺的直边画另外一条线，与刚画的弧线呈精确的90度角。画这条线，测量你想要的领宽。这将是服装上领子的后中心线。

步骤7：画领子边缘线。你可以修正这条线，得到你想要的领子的形状。

步骤8：向下测量直到前中心线，标记你希望的领口线的最低位置。你可以称这点为E。（在这个事例中，点E就在原前领口线的中心点上）。

步骤9：从B点通过E点画一条线。这条线与后领口线连接圆顺，并且通过前中心点继续延长几英寸。

步骤10：画领子的外边缘线与新的领口线相交。在衣身纸样上产生了一个缺口，领片与领口线相交时，缺口两边相等。

步骤11：从F点画一条弧线与止口点相交。

现在将这个图分解为几片。左前片和右前片，领面和底领。底领裁剪为2片，并使用斜丝缕面料（图13.4和图13.5）。

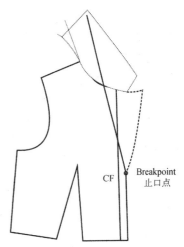

Figure 13.4 Dividing the pattern into separate pieces
图 13.4 将纸样分解为几片

 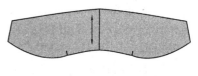

Figure 13.5 Upper and under collar
图 13.5 领面和底领

专业词汇

2. breakpoint *n.* 止口点

通用词汇

① extension *n.* 延长

Step 12: You will also need to draft a facing for your collar. Measure out 2" from the center front hem, and 2"out from your new neck shoulder point. Connect the lines to form your facing.

Note: You always want to taper (reduce) your undercollars and facings by ⅛" along the outer edge to ensure that they roll to the underside of the garment and are not visible from the right side.

Instead of reducing ⅛" from your undercollar pattern piece, you can also make your top collar ⅛" larger. Add ⅛" to each edge, excluding the back neck and center back seam. (Only along the collar edge).

The notched collar facing is tapered a little differently than a regular facing, as the center front edge and the collar edge are connected (Figure 13.6).

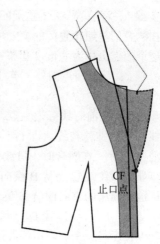

Figure 13.6 The notched collar facing
图 13.6 翻驳领挂面

Since the facing piece of a notched collar rolls outwards at the breakpoint and is seen as the right side of the garment, you will want to add ⅛" to the facing pattern piece along the collar edge, stopping at the breakpoint, so that the bodice front will roll to the underside. At the breakpoint the facing will be turned inwards and the front bodice will be seen as the right side once again. Along the center front, from the breakpoint to the hem, you will now taper the facing inwards by ⅛".

2. Drafting princess panel lines

Princess seam lines are a way of shaping bodice patterns to take into account the shaping around the bust and waist without the need for darts. In order for the seam line to do this effectively it has to pass over the bust point, or pass very close to the bust point. The further away from the bust point that you place your panel line, the harder it will be to shape the pattern pieces effectively without the need for additional darts or seam lines.

These seamline shapes are often then reflected in the seamlines of the back pattern pieces, so that on both front and back pattern pieces the panels curve into the armhole, or both panels meet at the shoulder.

Below are the instructions and diagrams for how to move the dart value from a waist dart to a princess panel line from armhole to waistline. This has been done using the "cut and spread" method of pattern making.

Trace off your pattern onto a new piece of card or paper so that you will not damage your original pattern.The basic bodice block (Figure 13.7) forms the starting point for the pattern manipulation.

Next draw on the guideline for your panel seam by drawing through the bust point and along the waist dart, then continue the guideline up from bust point into the armhole. Make sure to mark notch

步骤12：你还需要画领子的挂面。在底边前中线量出2英寸宽，颈肩点也量出2英寸。连接这两点，得到挂面。

注意：你需要在底领和里襟外边缘修剪掉1/8英寸，确保它们转到服装的里面，从正面看不到。

除了底领外边缘减少1/8英寸外，你还可以使领面增大1/8英寸。后领口线和后领中心缝份线不可以增大，每一边缘增大1/8英寸（只是沿着领口的边缘增大）。

翻驳领的里襟上大下小，与正常里襟有点不一样，因为前中心边缘和领子边缘相连。

因为翻驳领的里襟在止口点朝外翻转，在服装正面被看到，因此在里襟纸样沿领口边缘到止口处应该增加1/8英寸，使前衣身朝里翻转。在止口点，里襟朝里翻转，重新看到前衣片正面。沿前中心线，从止口点到底边，里襟收进1/8英寸。

2. 画公主线

公主线是修整上衣身合体的一种方法，它围绕胸部和腰部修整，但不需要收省。为了有效地使用这种方法，缝线通过胸高点，或接近胸高点。你设置的缝线越是远离胸高点，并且不添加额外省道或缝线想有效地获得纸样衣片就越困难。

这些缝线形状经常反射到后衣片上，因此前后衣片的弧线汇入袖窿，或它们在肩部相交。

下面介绍和图示如何将腰省转移到从袖窿到腰线的公主线的衣片。这是一直使用的"剪和展"的纸样裁剪方法。

拓印你的纸样到一张新的卡纸或纸上，这样不会损害你原来的纸样。基础原型纸样（图13.7）是操作纸样第一步。

下面在衣片上通过胸高点，并沿着腰省，画一条引导线。然后继续向上从胸高点到袖窿

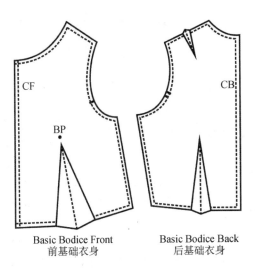

Figure 13.7　The basic bodice pattern
图 13.7　基础衣身原型

points for bust point and 1" above and below this bust point. These notches will help the curve to be sewn correctly (Figure 13.8).

Cut the pattern along the guidelines, discarding the dart value. Design lines are often drawn through the highest and lowest curves of the body for close fitting garments, such as the bust point, or around the waist. This enables the panels to be cut in as close to the body as possible.

Place the patterns onto a new sheet of cardboard. Trace the outlines of the patterns onto the cardboard underneath, and smooth the curve of the panel lines by hand, by using a french curve ruler[3], or by using a similar pattern making guide (Figure 13.9).

Make sure that the seam lines of the princess panel lines are equal to each other in length. Draw the seam allowance onto the new panel lines and transfer the notch marks (Figure 13.10).

Cut out the pattern pieces making sure that all markings have been transferred. The grainline will run parallel to center front for the front bodice piece. The grainline for the side panel will run up the center of the panel, perpendicular to the hemline—assuming that the hemline represents a horizontal waistline (Figure 13.11).

Note that the princess line panel may require more shaping in the waist section, from the waist hem line up to the first notch.

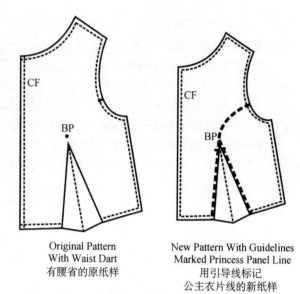

Figure 13.8 Original and new pattern with guidelines
图 13.8 原纸样和画有引导线的新纸样

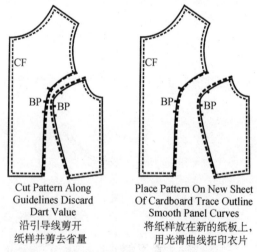

Figure 13.9 Cutting and tracing pattern
图 13.9 剪开和拓印纸样

It is always recommended that a toile version of a new pattern be sewn up to test the pattern and the fit. You may find that you want to adjust the curve of the princess line by eye by drawing a new line onto the test toile, and then making the adjustment on the pattern.

3. Drafting basic yoke line

When patterns are fairly simple, it is usually quite straight forward how individual pattern pieces

画线。在胸高点和胸高点上下1英寸处打剪口。这些剪口将有助于准确地缝合弧线（图13.8）。

沿着引导线剪开纸样，舍弃省量。设计线经常通过身体上最凸和最凹的曲面上画线，这样能使服装更加合体，例如胸高点或腰部周围。这样尽可能使衣片与身体吻合。

将纸样放到一张新的纸板上。拓印纸样的轮廓线到下方的纸板上，用手画、或用法线弧线尺或类似的纸样制作辅助工具将曲线画圆顺（图13.9）。

确定公主衣片的缝线彼此等长。在新衣片上画出缝份，并打剪口（图13.10）。

剪切纸样衣片，确定标上所有剪口记号。前衣片上的丝缕线平行于前中线。侧衣片的丝缕线画在衣片的中央，与底边线垂直——假定底边代表水平的腰线（图13.11）。

注意公主衣片线在腰间部位，即从腰底边向上到第一个缺口，需要很多修整。建议新的纸样先用坯布缝制做合体试验。通过眼睛观察，你可能找到你想要修正的公主弧线，在坯布上画新的线，然后在纸样上作相应调整。

3. 画基础育克线

当纸样相当简单的时候，通常很直接地由每一个衣片来命名。例如，一件简单的

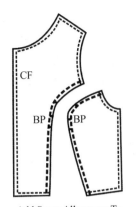

Add Seam Allowance To The Panel Lines. Ensure That Seamlines At The Same Length And Notches Match
在剪开衣片边缘添加缝份，并确保缝线长度相等和刀眼对齐

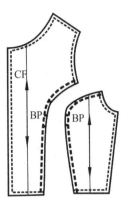

Cut Out Final Patterns Final Grain Lines Run Parallel To CF And Through The Center Of the Side Panels
剪裁最终纸样，丝缕线最终与前中线平行和贯穿于侧衣片的中间

Figure 13.10　Add seam and grainlines
图13.10　添加缝份和丝缕线方向

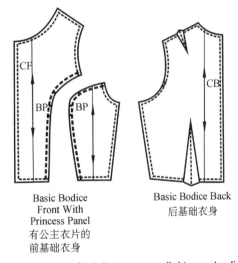

Basic Bodice Front With Princess Panel
有公主衣片的前基础衣身

Basic Bodice Back
后基础衣身

Figure 13.11　Grainline run parallel to center line
图13.11　直丝缕线平行于中心线

专业词汇

3. french curve ruler *n.* 法线弧线尺

need to be named. For example on a simple t-shirt you basically just have a front, a back and a sleeve pattern. For a jacket you may have front, back, sleeve, pocket pieces, facings, collars and linings. All of which still relate obviously fairly to parts of the garment. But there is one garment area with a more unusual name, and that is what is known as a yoke.

On garments, a yoke is probably most commonly seen on men's shirts, but also appears on some tops, dresses, trousers, skirts and coats. It is effectively the name given to a horizontal panel near the shoulders or waist and is a piece that is often used for shaping, since dart values can sometimes be absorbed into this seam line.

Yokes can be used as a design detail, but also create an opportunity for the pattern maker to better fit the panels of the garment to the body without darts. For example, a yoke panel is often used in jeans to allow the fabric pieces to mould[②] better around the curves of the body.

Depending on the shape of shoulder line required, yokes on tops, shirts, dresses or coats can sometimes be cut as one pattern piece so that the front and back pattern pieces are merged along the shoulder line. Other variations will keep the front and back yokes as two separate pieces, especially where there is a drop shoulder as you will need two separate pieces in order to retain the curve of the shoulder line. A variation on this would be to cut the yoke in one piece but with darts in the shoulder to create the necessary curved shaping. Yokes do not have to be straight across the horizontal lines for the front or back pattern, and in some of the variations that the yokes come to a point。

On trousers and skirts the yoke section is a way for the dart value to be absorbed into a single panel, but is similar to the technique used in Figure 13.12 for folding out the back shoulder dart. Below are some basic examples of pattern manipulations for different basic yoke patterns on a basic women's bodice block.

1) Draw a line on basic bodice of front and back (Figure 13.12).
2) Split the line to add yoke seamline (Figure 13.13).

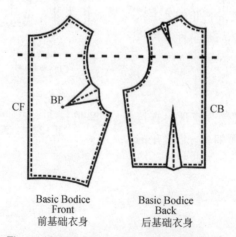

Figure 13.12　Draw a line on basic bodices
图 13.12　在基础原型衣上画一条直线

T恤，总体上说有一片前片、一片后片和一片袖片。一件夹克，有前片、后片、袖片、袋片、里襟、衣领和里子纸样。所有这些都非常明显地与服装的某个部分有关。但是，服装上的一个区域有一个非同寻常的名字，它就是我们知道的育克。

在服装上，最常见男性衬衫上的育克，但它也会出现在一些上衣、连衣裙、裤子、裙子和外套上。实际上这个名字是指靠近肩部、腰部的水平衣片，这个衣片经常用于造型，因为有时省量在这条线上被收进。

育克可以用作设计细节，同时也给纸样师一个创造性机会，使服装更加合体而没有省道。例如，育克衣片经常用于牛仔裤，衣片更好地吻合身体曲线。

根据想要的肩线形状，上衣、衬衫、连衣裙或外套的育克有时可以裁剪为一片衣片，使前后衣片纸样沿着肩线连成一体。还有一些变化是使前片和后片育克分开，特别是落肩的时候，两片分开是为了保持肩部的弧线。这种情况下，有一种变化是将育克裁成一片，但是在肩部有省道，产生必要的曲线形状。在前后衣片上育克不是非得处理成水平状态，在一些变化中育克线汇聚到一点。

在裤子和裙子上，育克部分是吸收省量成为单个衣片的一种方法，它类似于图13.12中折叠肩省的方法。下面是在女装基础衣身原型上进行的一些基本操作方法，可以用于不同的基础育克纸样上。

1）在前后基础原型衣片上画一条直线（图13.12）。
2）剪开这条线，即添加育克缝线（图13.13）。

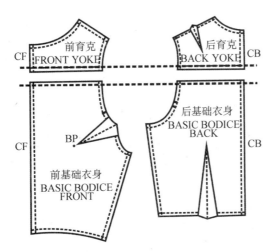

Figure 13.13　Add yoke seamline
图 13.13　添加育克缝线

通用词汇

② mould *vt.* 用模子做

3) An example of front yoke being double along centre front line to form one single yoke piece. Remind that CF line in Figure 13.13 was seam line, not on outside edge (Figure 13.14).

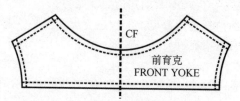

Figure 13.14　One single front yoke piece
图 13.14　一片式前育克

4) An example of back yoke having dart value folded out finished shoulder seam line will need to match shoulder seam of front yoke. Seam will need to be smoothed (Figure 13.15).

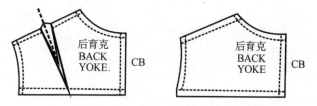

Figure 13.15　Folding out dart value
图 13.15　折叠省量

5) An example of joining the front and back yoke to remove the shoulder seam. Seams and curves of neckline and armhole will need to checked and smoothed (Figure 13.16).

3）举例，沿着前中线折叠，形成一片式前育克。注意前中线是图13.13中的缝线，而不是边缘线（图13.14）。

4）举例，后肩省折叠后产生新的肩线，并形成后育克，但是后肩线必须与前育克肩线吻合。缝线要画光滑（图13.15）。

5）举例，前后育克连接，取消肩线。缝线和领口弧线以及袖窿弧线需要检查和画光滑（图13.16）。

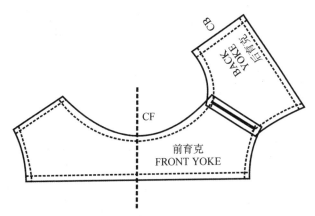

Figure 13.16　Joining the front and back yoke
图 13.16　连接前后育克

Passage 14

Basic Sewing Techniques of Jackets

Being familiar with terms used in menswear tailoring will help you set the collar and accurately stitch the facings. The roll or breakline[1] referred to in women's tailoring becomes the creaseline[2] in men's tailoring (Figure14.1), where the lapel and collar turn back. The line is a soft roll in women's tailoring and a definite crease in menswear.

The creaseline of the lapel usually begins just above the top button. The number of buttons on a jacket, however, may affect the beginning of the creaseline. For example, a three-button jacket is buttoned on the middle button. Check the pattern before you begin to see where the creaseline falls.

The collar stand lays between the neckline and the creaseline. The fall doubles down over the stand when a garment is worn (Figure14.2). The fall should be at least 3/8" wider than the stand. The fall also should cover the neckline seam of the garment by 3/8" to 1/2" when finished.

Another tern used in menswear is the gorgeline[3], its portion lays between the point where the creaseline meets the neckline and the point where the collar joins the lapel (Figure14.3).

The collar of a man's jacket hugs the neckline and lays flat across the back. This compares to women's collars that are often planned to stand away from the neck and lay loosely across the shoulders. The menswear lapel lays flat against the body while women's wear lapels are rolled to suit body contours. Special techniques assure the closer fit of the man's jacket.

1. Undercollar

The undercollar used in menswear tailoring is cut from a

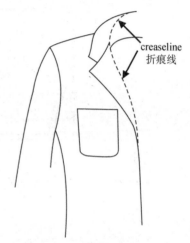

Figure 14.1 Creaseline of collar and lapel
图 14.1 折痕线和驳头

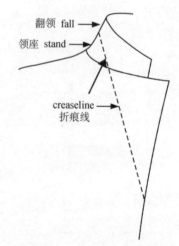

Figure 14.2 Collar stand and fall
图 14.2 领子上的站领和翻领

第 14 课

上装基础缝制工艺

了解缝纫男装时使用的术语,有助于你安装领子和精确地缝制挂面。在女装中,翻转线或翻折线在男装中称为折痕线,驳头和领子在折痕线上翻折(图14.1)。在女装中,它是一条柔软的翻转线,在男性中是一条明确的折痕。

驳头的折痕线通常从第一粒纽扣开始。但是,上装纽扣的数量可能会影响折痕线的起始。例如,三粒扣的夹克扣中间一粒扣。在你决定折痕线的起点之前,先查看纸样。

领座在领口线和折痕线之间。当服装穿在人体上的时候,翻领翻下盖过领座(图14.2)。翻领部分必须比领座部分至少宽3/8英寸。成品服装中,翻领要盖过服装的领口线3/8~1/2英寸。

男装中的另一个术语是串口线,它在折痕线和领口线的交点与衣领和驳头的交点之间(图14.3)。

男装衣领紧裹领口线,在颈后呈平坦状。比较女装的衣领,经常设计为竖立并离开颈部,从肩部向后很宽松。在身体上男装的驳头平坦,而女装驳头翻下后与身体曲线吻合。男性夹克的独特技术就是更加合体。

1. 底领

男装底领的缝制,要求裁剪一块

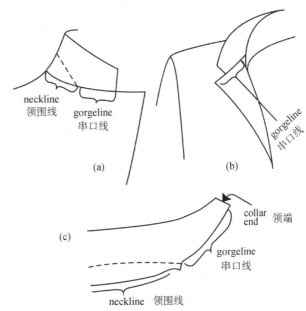

Figure 14.3 Gorgeline of Jacket
图 14.3 夹克的串口线

专业词汇

1. breakline *n.* 翻折线　　　　2. creaseline *n.* 折痕线　　　　3. gorgeline *n.* 串口线

fabric that will allow a sharp crease in the collar. The edges are usually left raw and the undercollar is fastened to the upper collar by a special hand or machine stitch called a fell stitch[4], another name for hem stitch[5]. A whip stitch[6] or overhand stitch[7] also may be used (Figure14.4).

Melton[8] cloth is used by manufacturers for undercollars, but this material is rarely available to the individual who tailors at home. Excellent substitutes are a good grade of wool flannel[9], cotton felt[10], a blend of rayon[11] and wool felt or a closely woven wool fabric that will not fray①. Self-fabric may be used if a sharp crease can be made in the undercollar. If fraying may be a problem, edges can be finished before the collar is applied.

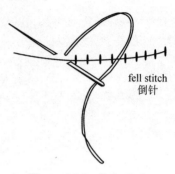

Figure 14.4   Fell stitch
图 14.4   倒针

The undercollar must be cut on the bias. If possible, eliminate the seam in the pattern and place the center back on a true bias fold. This will be an aid in achieving a close fitting collar. Sometimes there is not enough fabric to allow the undercollar to be cut without a seam. In fabric that presses well into a flat seam, the center back seam can be stitched and pressed as usual. If fabric does not press easily into a flat seam, trim off the seam allowance and center the cut edges over a strip of seamtape or muslin. Zigzag over the raw edges. Self-fabric that frays will have to have a seam in the center back of the undercollar.

Non-fraying fabric. Trim the seam allowance to ¼" on the neckline and gorgeline. Trim all of it off the other edges. Though this may be done as the collar is cut, accuracy will be increased if the collar is trimmed after it is cut (Figure14.5).

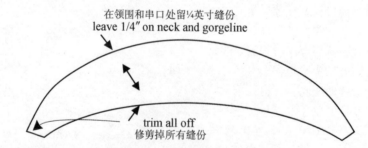

Figure 14.5   non-fraying fabric- trim seam allowance on outer edge of collar
图 14.5   不脱丝面料——在领子的外边缘修剪缝份

Fraying self-fabric. Trim seam allowances to 3/8" and mark the seamline with a hand or machine basting stitch (Figure14.6).

2. Interfacing[12] for the Undercollar

Canvas without hair fiber makes the most satisfactory interfacing for undercollars because it

面料，上面有清晰的折痕。边缘通常留着毛边，并且用特别的手工针迹或机器针迹钉在领面上，这种针迹称为倒针，另一个名称叫卷边针。也可以使用鞭子针或锁边针（图14.4）。

麦尔登面料被制造商用来制作底领，但是家庭的个体裁缝很难弄到这种面料。最好的替代品是优质羊毛法兰绒、棉毡、人造丝和羊毛毡的混纺织物，或一种不易磨损的紧密织物。与服装相同面料，如果能够制作出清晰的折痕也可以使用。如果布边缘容易脱丝，可以在用于衣领前先拷边。

底领必须斜裁。如果可能，去掉纸样上后中缝的缝份，将后中缝对齐正斜对折的面料上，这样有利于得到合体的领子。有时没有足够的面料可以使底领中线没有接缝。那就选择一种能够烫平接缝的面料，像往常一样拼接和熨烫。如果面料不能轻易地烫平接缝，那就剪去缝份，将后中线对齐放到一根缝接带子或平纹细布上，在接缝上采用锯齿型针迹缝合毛边。与服装相同的、易于脱丝的面料，在底领后中心不得不有一条接缝。

不脱丝的面料。修剪领口线和串口线，保留¼英寸缝份。剪去其他边缘的缝份。尽管在剪裁领子的时候可能已经修剪了，但是如果剪裁以后再修剪一下，精确度会提高（图14.5）。

与服装相同的易于脱丝的面料。修剪缝份，留有3/8英寸，用手工或机器粗缝标记缝线（图14.6）。

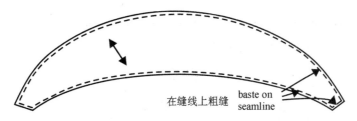

Figure 14.6　Fraying fabric 3/8" seam allowance
图14.6　脱丝面料留有 ³⁄₈ 英寸缝份

2. 底领内衬

不含毛纤维的帆布最适合做底领内衬，因为它可以有很清晰的折叠线，而不是翻转线。亚麻帆布是最理想的，但是不容易找到。在这种情况下，人造丝和棉混纺的帆布与亚麻的功

专业词汇

4. fell stitch *n.* 倒针
5. hem stitch *n.* 暗卷针
6. whip stitch *n.* 鞭缝针
7. overhand stitch *n.* 锁边针
8. melton *n.* 麦尔登呢
9. flannel *n.* 法兰绒
10. felt *n.* 毛毡
11. rayon *n.* 人造丝
12. interfacing *n.* 内衬

通用词汇

① fray *vt.* 磨损

creases sharply rather than rolls. Linen canvas is ideal, but it is also difficult to find. Canvas made of a combination of rayon and cotton works as well as the linen in this situation. If the canvas you are using for the front interfacing is of a cotton-rayon blend with a small percentage of goat hair, you could use this in the undercollar.

The interfacing should also be cut on a bias fold if possible. If not, the raw edges of the center back may be overlapped and stitched along the seam allowance, then trimmed close to the stitching (Figure14.7).

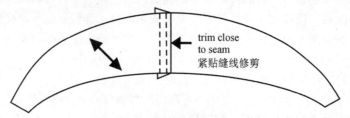

Figure 14.7　Overlap CB–trim seam
图 14.7　后中心线重叠，修剪缝份

Center back seam allowances may be trimmed off and the cut edges centered over a strip of seam tape or muslin and zigzagged into place (Figure14.8).

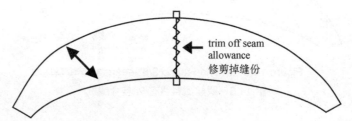

Figure 14.8　Center tape under edge
图 14.8　后中心线下衬一带子

If there is a center back seam taken in the undercollar, the seam allowances plus ¼" may be trimmed from the center back of the interfacing and slipped under the center back seam allowances of the undercollar. Catchstitch[13] it along the undercollar seam (Figure14.9).

Trim the seam allowance from the outer edges of the interfacing. Greater accuracy will be gained if the trimming is done after the interfacing for the undercollar is cut out (Figure14.10).

After the interfacing is cut and ready to use, lay it on the undercollar. Press both on a flat surface. Check to be sure enough has been trimmed from the interfacing before proceeding.

Fusible[2] interfacings may also be used. Trim off the seam allowances plus an extra ¼" from the fusibles before applying them to the undercollar (Figure14.11).

Experiment with scraps[3] first because some iron-ons will not press into a sharp crease nor allow enough stretch and shrinkage[4] in portions of the undercollar. Fusing webs can be used with a

能差不多。如果你使用帆布作前衣片的内衬，那就用棉和人造丝混纺面料，并含有少量比例的山羊毛，你也可以在底领上使用这种内衬。

内衬要尽可能斜向折叠裁剪。如果不是这样，后中线毛边可以重叠，沿着缝份缝合，然后紧靠缝线修剪（图14.7）。

可以剪去后中线缝份，将毛边放在缝带或平纹细布上，Z型针缝合（图14.8）。

如果底领后中心线有缝合的缝，内衬后中线除了要剪去缝份，还要再剪去¼英寸，使内衬的后中心边缘钳在底领的后中心缝份内。沿着底领缝份边缘缝叠针针迹（图14.9）。

修剪内衬的外侧边缘。在底领剪好以后，修剪内衬将更加精确（图14.10）。

内衬修剪以后等待使用，将它放置在底领上，同时熨平整。在做下一步之前检查内衬修剪得是否到位。

也可以使用可熔性内衬。将它用于底领之前，除了修剪熔衬的缝份外，还要再修剪掉¼英寸（图14.11）。

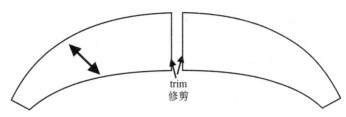

Figure 14.9　Trim off seam allowance
图 14.9　剪去缝份

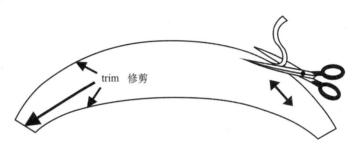

Figure 14.10　Trim outside seam allowance
图 14.10　修剪外侧边缘缝份

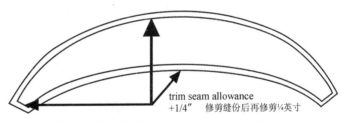

Figure 14.11　Trim fusibles more than non-fusible interfacing
图 14.11　热熔内衬比非热熔内衬修剪的量多

先用余料做试验，因为一些热熔衬不能烫出清晰的折痕，在底领上也不能有足够的伸缩性。热熔网可以和薄型梭织面料结合在一起做内衬。也要用余料做试验，看看是否能够取得

专业词汇

13. catchstitch *n.* 叠针

通用词汇

② fusible *adj.* 可熔的　　　　③ scrap *n.* 废料　　　　④ shrinkage *n.* 收缩

lightweight woven interfacing. Again, experiment with scraps to see if you are gaining the desired effect. The interfacing seam allowances plus 1/4" are trimmed off before fusing. Fusing webs should be trimmed just short of the interfacing.

Mark the creaseline, notches and shoulder line with pencil or a tracing wheel on the interfacing. Also mark the center back of the undercollar if there is no seam (Figure 14.12).

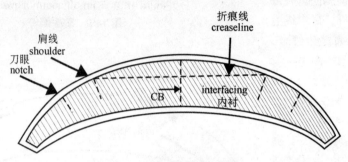

Figure 14.12　Mark interfacing
图 14.12　在内衬上做标记

Transfer the markings of all but the creaseline to the right side of the undercollar with hand or machine basted marks with contrasting color thread (Figure14.13). Do not use clips to mark construction points since the edges will not be enclosed in a seam.

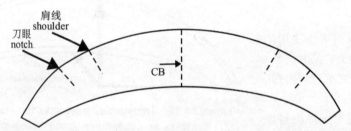

Figure 14.13　Transfer markings to right side of the undercollar
图 14.13　转移标记点到底领的正面

For self-fabric undercollar that frays. Catchstitch a nonfusible interfacing to the undercollar (Figure14.14).

Miter the corners of the undercollar and catchstitch the raw edge of the undercollar to the interfacing, and finish edges of a fraying fabric (Figure14.15).

3. Establish the creaseline

The creaseline may be established several ways. It may be held in place by temporary hand or machine basting stitches. It may also be established permanently by a machine-stitched line, using the regular size stitch or a hand backstitch (Figure 14.16).

理想的效果。在热熔之前除了修剪内衬的缝份，还要再修剪掉¼英寸。热熔网应该刚好短于内衬。

用铅笔或拓印滚轮在内衬上标记折痕线、刀眼和肩线。如果底领上没有后中缝，还要标记后中心点（图14.12）。

除了折痕线，转移其他所有标记点到底领的正面，用对比色线，手针或机器粗缝（图14.13）。不要用打剪口的方法标记这些结构点，因为它们在接缝里不能闭合。

易于脱丝与服装面料相同的底领。将非热熔内衬用叠针缝到底领上（图14.14）。

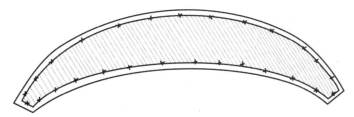

Figure 14.14　Catch stitch interfacing to undercollar
图14.14　用叠针法将内衬缝到底领上

斜接底领领角，用叠针法将底领的毛边边缘缝到内衬上，完成底领毛边处理（图14.15）。

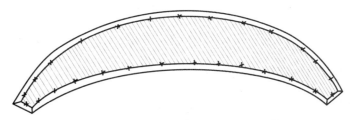

Figure 14.15　Miter corners of undercollar
图14.15　缝合底领边缘

3. 确定折痕线

有几种确定折痕线的方法。可以用手针或机器粗缝临时的线迹。也可以用机器，以正常针脚或手工缝后退针迹，永久性缝合（图14.16）。

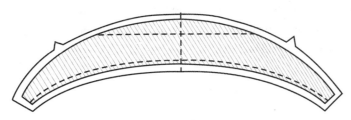

Figure 14.16　Establish the creaseline
图14.16　确定折痕线

Use matching color thread. Tape is not usually used on undercollars of man-tailored jackets. If you want to use it, place it below the creaseline in the stand area, so that it will join the line of the tape from the lapel. Place it along the edge of the creaseline rather than centered on the creaseline.

Press in the Creaseline. Lay the collar flat with the stand uppermost. Press with steam, slightly shrinking the creaseline so that it will fit snugly against the neck. Beginning at the center back and moving to the ends of the collar, slightly stretch the outside edges of the fall. This will put a curve into the collar (Figure 14.17). Be careful not to shrink or to stretch the collar from the original proportions. Remember that you are shaping the collar to the neckline without changing its lines. Let the undercollar dry completely before working with it further.

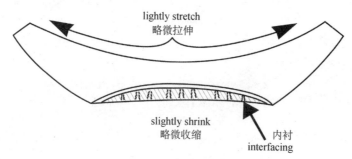

Figure 14.17 Press creaseline
图 14.17 熨烫折痕线

4. Test for Fit on the Garment

Pin the undercollar to the neckline of the jacket, carefully matching center back and construction marks. Lap the stand edge of the undercollar over the seam line of the garment neckline (Figure 14.18). If all points match perfectly, you are ready to pad stitch[14].

If the construction points do not match and the neckline has stretched rather than shrunk, you may need to reshrink or recut this edge so that it does fit. Recut the whole collar if it is too much out of shape. Check at this time to be sure that the fall comes at least 3/8" below the neckline seam.

5. Stitching undercollar

Pad stitch the stand. Pad stitch from the creaseline to the edge of the interfacing (Figure14.19). The rows of stitches may be vertical or parallel to the creaseline, but the vertical rows will give a firmer, stiffer stand. Stitches should be small and the rows about ¼ apart. Begin at the center back and

---

专业词汇

14. pad stitch *n.* 纳针

使用色彩匹配的线。男性夹克上的底领上一般不用粘带。如果你想使用，将它放在折痕线下方的领座区域，使它与驳头的黏带连接起来，并沿着折痕线的边缘而不在折痕线中间。

熨烫折痕线。将领放平，领座在上方。带蒸汽熨烫，微微收缩折痕线，使它能够贴近脖颈。从后中心开始，然后移向领子的两端，微微拉伸翻领的外边缘，使衣领呈弧线（图14.17）。小心不要收缩或拉伸超出原来的比例。记得在你对领子的领围线进行造型的时候，不要改变边缘线。在底领完全干了以后，再继续操作下一步。

4. 底领试样

用大头针将底领别合到夹克的领围线上，仔细对齐后中心和结构标记点。将底领领座的边缘覆盖到服装领围线上。所有的点都完全对好后，就可以用扳针法将它们缝合在一起（图14.18）。

如果结构点不吻合，领口线拉伸太多而不是收缩，你需要重新收缩或重新裁剪这个边缘，使得它们吻合。如果变形太多，就要重新裁剪整个领子。此时要确定翻领至少覆盖领围线3/8英寸。

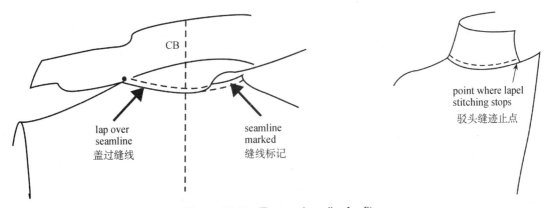

Figure 14.18　Test undercollar for fit
图 14.18　试验底领

5. 纳底领

纳针法缝领座。从折痕线到内衬边缘缝纳针（图14.19）。每排针迹可以垂直或平行于折痕线，但是垂直排列使领座更加结实和硬挺。针迹必须很小，每排之间的距离大约¼英寸。

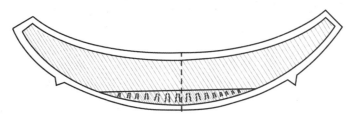

Figure 14.19　Pad stitch the stand
图 14.19　纳针缝领座

work outward. Hold the neckline edge toward you. Stop the stitches at the edge of the interfacing. Be sure not to pull stitches too tight or to let them show on the right side.

Pad Stitch the Fall. Beginning at the creaseline, pad stitch the fall with rows parallel to the creaseline (Figure 14.20). These rows should be farther apart and the stitches should be longer. Hold the edge of the collar toward you.

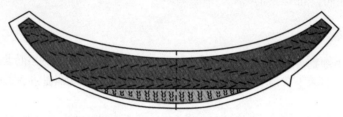

Figure 14.20　Pad stitch the fall
图 14.20　纳翻领

Do not pad stitch outside the interfacing. Begin at the center back for each collar half to prevent pulling stitches too taut[5] or getting a lopsided[6] effect.

6. Shaping of the undercollar

Place the undercollar around a curved area such as a tailor's ham or a turkish towel rolled into a ball. Place it in the same position as it will be when worn. Use pins to hold it in place. Press the creaseline in with steam. Beginning at the center back again, gently stretch the outside edge of the fall with your fingers, especially through the areas going over the shoulder (Figure14.21). Use caution so that you do not pull the undercollar out of shape. Allow the undercollar to dry on the curved shape. Before the undercollar is applied permanently to the garment, the upper collar is usually prepared. Though the upper collar can be molded to the undercollar after the undercollar is attached, there is less bulk to handle if it is molded in the next step.

7. Upper collar

The upper collar is cut according to the pattern and corners are mitered. Mark the seamlines with hand or machine basting stitches. The center back and gorgeline may be marked with clips into the seam allowance. The marking on the neckline center back and gorgeline is an aid to accuracy when the collar is stitched to the garment. The marking on the outside seamlines is an aid to accuracy when corners are mitered.

While the undercollar is still on the curved form, match corresponding construction points on the outside edges of the fall. Wrong sides will be together. Use pins to hold them in place. Fold the stand area of the upper collar under at the crease line so that it corresponds to the creaseline of the undercollar. The stand area will be sandwiched between the upper collar and the undercollar (Figure14.22). Use pins to hold. Steam press to mold the upper collar around the undercollar.

从后中线开始，然后向外拓展。缝制时领线边缘朝向自己。在内衬边缘停止缝针。缝线不要拽得太紧，也不要在正面看到针脚。

纳针法缝翻领。从折痕线开始，然后一排排地平行于折痕线纳针（图14.20）。排与排之间分离得远些，针脚也长些。领的边缘朝向自己。

纳针不要超出内衬。从中间开始向两边缝，避免线拉得太紧或两边不均衡。

6. 底领塑形

将底领围绕到一个弧线形状的事物上，例如裁缝的烫包或用土耳其毛巾卷成的球，弧形与穿在人体上的弧线大致相同。用大头针固定它。用蒸汽熨斗熨烫折痕线。从后中心开始，轻轻地用手指拉伸翻领的边缘，特别是经过肩部的那个区域（图14.21）。谨慎地，不要将底领拉走形。保持底领呈弧形待干，在永久地安装到服装上之前，准备好领面。领面可以在底领与其缝合后塑形，但是如果按照下面的步骤塑形，操作会容易些。

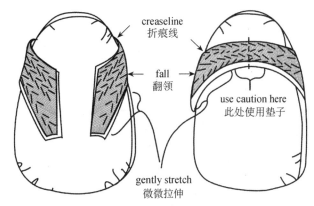

Figure 14.21　Final shaping of the undercollar
图14.21　底领最终的形状

7. 领面

领面根据纸样裁剪，并封领角。用手工或机器粗缝标记缝线。后中线和串口线可以用剪口在缝份上做标记。标记领口线中点和串口线，有助于准确将领子缝到服装上。当领角封口后，标记外侧缝线有助于精确性。

当底领仍然处于弧线形状的时候，将相应的结构点与翻领外边缘对应，反面对反面，使用大头针将它们别合在一起。在折痕线下方折叠领面的领座区域，使它与底领的折痕线一致。领座区域夹在领面和底领之间（图14.22）。用大头针固定住。用蒸汽熨烫，围绕底领塑形领面。

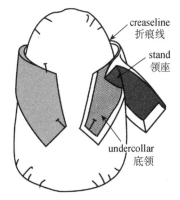

Figure 14.22　Mold uppercollar
图14.22　领面整形

通用词汇

⑤ taut *adj.* 紧的　　　　　　⑥ lopsided *adj.* 两侧不匀称

Upper collars are usually cut a bit larger than undercollars to allow for needed give in the upper collar. Allow for this and do not try to shrink the collar or mold it too tightly to the undercollar. Allow the collar to dry and lay it aside so that the shaping will be retained until needed. The upper collar is stitched to the garment after the facings and undercollar are applied.

Overlap the 1/4" seam allowance left on the neckline and gorgeline edges of the undercollar over the seamline of the garment. Baste the collar into place close to the edge. The permanent fell stitch will be put in after the uppercollar is in place.

Try the jacket on the individual to check for fit. Make any needed adjustments to assure that the undercollar hugs the neck and lays flat before proceeding.

8. Front facings

Accuracy is important when the facings are marked for construction points and when these points are matched to the body of the garment. Carefully mark the slash line where the stitching ends on the lapel. Most facing patterns allow for slight ease where joined to the garment, and the pattern piece will be slightly larger through this area then the jacket front.

The front and back linings are stitched to the front facing along the shoulder line before the facing is stitched to the garment (Figure14.23). There is no back facing as is usual in women's garments. The lining sleeves are put in by hand later with a fell stitch. The underarm seams of the lining are stitched by hand or machine after the upper collar is stitched on.

Turn the garment right side out. Place the facing on the garment, right sides together. Beginning at the hem edge, match construction points up to the slash mark at the end of the lapel. Pin. Baste if necessary.

Beginning at the hem edge, stitch the facing up to the slash point. Backstitch at each end. Be sure to ease the facing onto the garment and slightly stretch the garment lapel as you stitch. Take a diagonal stitch across the point of the lapel, rather than pivoting the needle at the point to give a sharper corner when turned.

Slash at the denoted point at the end of the stitching line. Trim and grade the seams around the lapel point to about 1/4" (Figure14.24). If interfacing has been stitched into the seam, trim it as closely to the seam as possible.

The seams down the front are graded so that the widest layer will lay next to the portion that shows on the outside of the garment (Figure14.24). The lapel will be graded so that the widest area will lay next to the facing. At the point where the facing reverses from the outside to the underside, clip into the seam and reverse the grading. This will keep the widest seam allowance on the top and prevent a ridge from showing on the right side. Usually 3/8" for the wider seam allowance and 1/4" for the narrower are sufficient. Press seams open, then to one side. Those in the lapel area will turn toward the garment, the portion below the lapel will turn toward the facing.

领面通常比底领略微大些，这是领面所需要的。考虑到这点，就不要尝试收缩领子，或者在底领上塑形得太紧。待领子干了以后，将它放在一旁，保持形状，直到需要它的时候。在挂面和底领缝好后，领面与服装缝合。

底领的领口线和串口线与服装的领围线重叠¼英寸，紧靠边缘粗缝领子，在领面归位后，使用回针永久地缝合。

将夹克放在人体上检查是否合身。在需要的地方作修正，确保底领包住脖颈，并且放置平复后再继续。

8. 前挂面

当挂面标记了结构点，并且这些点与服装衣身吻合的时候，重要的是精确性。驳头缝线末端要细心标记剪口。通常挂面与服装衣片缝合的时候，要有微微的松量。在前衣身区域，挂面要比衣身纸样略微大些。

先将前后衣身的里子在肩缝对齐与前挂面缝合起来，然后才能将挂面与服装缝合（图14.23）。在女装中通常没有后挂面。袖里子稍后手工回针缝制。在领面缝制完成后，手工或机器缝制手臂下方缝。

将服装的正面翻出，挂面正面放在服装上正面相对。从底边边缘开始，对齐结构点向上直到驳头末端的剪口，用大头针别合。如果需要的话粗缝。

从底边开始，向上缝合挂面到剪口，在末端回针。挂面放在服装上时一定要有微微的松量。当缝合的时候，略微拉伸服装的驳头。在驳头的顶点缝对角线针，而不是以这点为轴旋转针脚，使驳头翻过来之后出现很尖的角。

在缝线末端标注的点处打剪口。在驳头点周围梯阶地修剪，保留大约¼英寸缝份（图14.24）。如果挂面已经被缝上，尽可能靠近缝线修剪。

前衣片向下的缝份要有梯阶，露在服装外面那个部分的缝份最宽（图14.24）。驳头部位也要有梯阶，最宽区域紧邻挂面。在挂面从外面翻转到里面的那一点，打剪口倒缝，翻转梯阶，使最宽的缝份在上方，避免缝脊显露在正面。通常较宽缝份是3/8英寸，较窄缝份是¼英寸。开缝熨烫，然后翻到一边。在驳头地方的缝份转向服装，在驳头下方的缝份转向挂面。

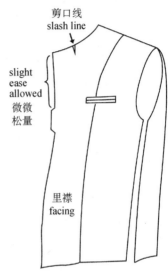

Figure 14.23 Front facing
图 14.23 前挂面

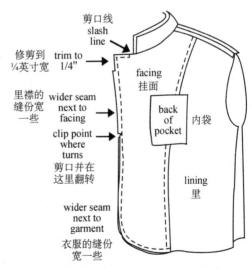

Figure 14.24 Trim and grade seams
图 14.24 修剪和梯阶缝份

9. Apply the upper collar

Right sides together, match construction points of the collar gorgeline and neckedge beginning at the slash mark. The finished edge of the upper collar will lay on the slash mark. Pin or baste the collar along the gorgeline of the facing and the neckline of the lining. Stitch along this seamline (Figure14.25). Backstitch at each end. The back of the garment with its attached undercollar will not be caught in this first stitching.

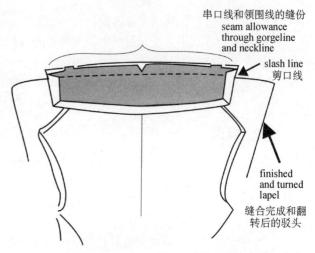

Figure 14.25 Stitch upper collar only to facing and lining
图 14.25 将领面与挂面和里子缝在一起

Turn the collar up and check to be sure that your stitching line begins exactly where the lapel stitching ended and that the collar was accurately placed (Figure14.26). You should have what looks like a continuous seam line from the right lapel point to the left one. Also check to be sure the edge of the undercollar falls just under the finished edge of the upper collar. There should be no bulges[7] where the two stitching lines meet and the collars begin their joining.

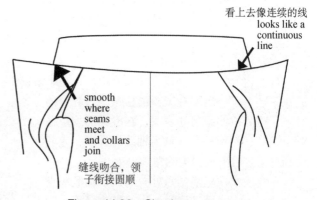

Figure 14.26 Check your accuracy
图 14.26 检查精确性

Notch into the seam allowance of the upper collar and facing through the gorgeline area to help the seam lay flat. Trim the seams to 3/8". Clip into the seam allowance where the lining joins the facing. Seams are pressed open through the gorgeline and pressed upward into the collar through the neckline (Figure 14.27).

Match center back lines of the garment and the lining. Trim the seams to 3/8". Stitch together with a basting stitch by hand or machine along the neckline (Figure14.28). The gorgeline seam of the facing is not fastened to the garment because give is needed as the garment is worn.

Match the upper collar and the undercollar. Pin or baste together. Fell the undercollar to the upper collar (Figure 14.29). The edge of the undercollar should fall just under the finished edge of the upper collar. Place the garment around a curved surface and give the collar a final press.

9. 安装领面

正面相对，对准两个结构点，即领串口线点和领围线打剪口的起始点。倒好缝的领面的边缘放在打剪口的这点上。沿着挂面的串口和里子的领口线与领面别合或粗缝（图14.25），再沿着这条缝线车缝，在末端缝回针。在这次车缝的时候，装有底领的服装后片将不缝合在一起。

将领面翻向正面，检查缝线是否准确地从驳头的缝线末端开始，领子是否安装准确（图14.26）。从右边的驳头点到左边的驳头点要看上去像一条连续的线条。还要检查底领的边缘在领面倒好毛边的边缘偏里一点。检查两条缝线相遇和领子开始缝合的地方没有臌胀。

领面的缝份打剪口，挂面在串口部位也打剪口，这样有助于缝份平服。修剪缝份留有3/8英寸。在里子与挂面缝合的地方修剪缝份。串口开缝烫平，在领线部位的缝份倒向领子熨烫（图14.27）。

对齐服装和里子的后中心线。修剪缝份留有3/8英寸。用手工或机器沿着领围线粗缝（图14.28）。挂面的串口线不要缝合到服装上，因为服装穿在人体上需要伸展性。

对齐领面和底领。别合或粗缝在一起。用回针将底领缝到领面上（图14.29）。缝好以后，底领的边缘比领面边缘里一点。将服装放置在弧形表面上，对衣领进行最终的熨烫。

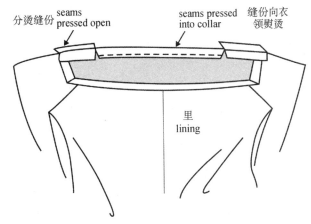

Figure 14.27 Press collar seams
图 14.27 熨烫领子的缝份

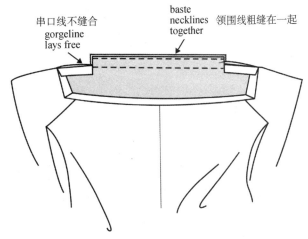

Figure 14.28 Stitch neckline of garment back to upper collar seam
图 14.28 将服装的后领围线与领面的领口线缝合

通用词汇

⑦ bulge *n.* 臌胀

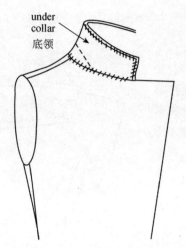

Figure 14.29　Fell collars together
图 14.29　回针缝合领面和底领

　　Accuracy in matching construction points and in exact measurement of seamlines is more important to applying the collar correctly than many years of sewing experience. Stop to check your accuracy after each step and do over what is inaccurate. This may seem tedious the first time, but with practice you will find that it goes quickly. You will also achieve the flat, neck hugging collar that is found in fine menswear tailoring.

结构点精确吻合，缝线尺寸精确，将这些技术正确地用于衣领上，比多年的缝纫经验更加重要。每一步骤完成后都要停下来检查是否精确，在不精确的地方作修改。第一次从事似乎很乏味，但是随着不断地实践，速度会提高。你将能够制作平复、包裹脖颈的衣领，与精致男装中的衣领相媲美。

Glossary 词汇表

A

accessory   *n.* 配饰
acetate   *n.* 醋酸人造丝
adjustment   *n.* 调整
adornment   *n.* 装饰
angle   *n.* 角度
ankle   *n.* 脚踝
apparel   *n.* 服装
applique   *n.* 缝饰
armhole   *n.* 袖窿
armour   *n.* 盔甲
armscye   *n.* 袖窿
asymmetrical   *adj.* 不对称
attire   *n.* 服装

B

bag   *n.* 包
balance   *n.* 平衡
ball-dress   *n.* 舞会服
baste   *vt.* 粗缝
batiste   *n.* 细棉布
batwing   *n.* 蝙蝠袖
bell-shaped   钟形
belt   *n.* 腰带
bespoke   *adj.* 全定制的
black-tie   男子半正式礼服
bleach   *vt.* 漂白
blouse   *n.* 女衬衫

bodice   *n.* 衣身
bondage   *n.* 绑带
brand   *n.* 品牌
brassiere   *n.* 文胸
breakline   *n.* 翻折线
breakpoint   *n.* 止口点
bricolage   *n.* 拼装
bright   *adj.* 亮色
buff   *n.* 米色
bust   *n.* 胸
buttock   *n.* 臀部
button   *n.* 纽扣
buttonhole   *n.* 扣眼
buyer   *n.* 买手

C

calico   *n.* 白棉布
cardigan   *n.* 开襟羊毛衫
cashmere   *n.* 山羊绒
catchstitch   *n.* 叠针
catwalk   *n.* 伸展台
chroma   *n.* 色度
closet   *n.* 衣橱
clothing   *n.* 服装
coat   *n.* 外套
codpiece   *n.* 遮阴布
collage   *vt.* 拼贴
collar   *n.* 衣领
colour   *n.* 色彩

contour　*vt.* 画轮廓
cool　*adj.* 冷色
corset　*n.* 紧身胸衣
cosmetic　*n.* 化妆品
costume　*n.* 戏装
cotton　*n.* 棉花
crease　*vt.* 折缝
creaseline　*n.* 折痕线
crisp　*adj.* 挺括
crossgrain　*n.* 横丝缕
crotch　*n.* 裤裆
curtain　*n.* 窗帘
curve　*n.* 弧线
custom-made　*adj.* 定制的

D

dart　*n.* 省道
design　*vt.* 设计
detail　*n.* 细节
diagonal　*n.* 对角线
dolman　*n.* 斗篷
draft　*vt.* 画纸样
draping　*v.* 立体裁剪
dresser　*n.* 衣着者
dressmaker　*n.* （女装）裁缝
dull　*adj.* 暗色
dummy　*n.* 人体模型
dusty　*adj.* 浅灰色
dye　*n.* 染料

E

earring　*n.* 耳环
ease　*n.* 松量
elbow　*n.* 肘部
ensemble　*n.* 全套服装
embellishment　*n.* 修饰

embroidery　*n.* 刺绣
eroticization　*n.* 性感
eveningwear　*n.* 晚装
excess　*n.* 多余量

F

fad　*n.* 狂潮
fashion　*n.* 时尚
fastening　*n.* 系扣
fell stitch　倒针
felt　*n.* 毛毡
fiber　*n.* 纤维
finger　*n.* 手指
finished edge　实边
flannel　*n.* 法兰绒
flat collar　平翻领
flesh　*n.* 肉体
flourish　*n.* 花样
form　*n.* 形式
formalwear　正式服装
fragrance　*n.* 香水
french curve ruler　法线弧线尺
fullness　*n.* 宽松度
fur　*n.* 毛皮
fuse　*vi.* 热熔

G

garb　*n.* 装扮
garment　*n.* 服装
gather　*n.* 碎褶
gender　*n.* 性别
girth　*n.* 周长
gorgeline　*n.* 串口线
grain line　丝缕线
grosgrain　*n.* 罗缎

H

hairline   *n.* 发际线
hairstyle   *n.* 发型
hanger   *n.* 衣架
hat   *n.* 帽子
haute couture   *n.* 高级时装
headdress   *n.* 头饰
heel   *n.* 后跟
hem   *n.* 底边
hem stitch   *n.* 暗卷针
hip   *n.* 臀
hipline   *n.* 臀围线
hosiery   *n.* 袜类
hue   *n.* 色相

I

illustrator   *n.* 插图画家
intensity   *n.* 纯度
inseam   *n.* 内长
interfacing   *n.* 内衬
iron   *vt.* 熨烫
inch   *n.* 英寸（1 英寸 = 2.54 cm）

J

jacket   *n.* 夹克
jean   *n.* 牛仔裤
jewelry   *n.* 首饰

K

kilt   *n.* 褶裥短裙
kimono   *n.* 和服
knee   *n.* 膝盖
knit   *n.* 针织

L

lapel collar   驳头领
layette   *n.* 婴儿的全套服装
leather   *n.* 皮革
leg   *n.* 腿部
light   *adj.* 淡色
linen   *n.* 亚麻布
lingerie   *n.* 贴身内衣
lining   *n.* 衬里
limb   *n.* 肢体
lip   *n.* 嘴唇
loom   *n.* 织布机
lower arm   前臂
luxury   *n.* 奢侈

M

make-up   化妆
made-to-measure   量体裁制
made-to-order   定做
mandarin   *n.* 中式领
mannequin   *n.* 人体模型
mask   *n.* 面具
masquerade   *n.* 化装舞会
match   *vi.* 匹配
mauve   *n.* 淡紫色
measurement   *n.* 度量
melton   *n.* 麦尔登呢
modify   *vt.* 修改
mouth   *n.* 嘴
muslin   *n.* 平纹细布

N

nail   *n.* 指甲
nakedness   *n.* 裸体
neck   *n.* 颈部

neckline　*n.* 领围线
neutral　*adj.* 中性色
notch　*n.* 刻痕
novelty　*n.* 新奇

O

obsession　*n.* 着魔
opaque　*adj.* 不透明
orange　*n.* 橙色
ornamentation　*n.* 装饰
outfit　*n.* 全套服装
outline　*n.* 轮廓线
overcoat　*n.* 大衣
overhand stitch　锁边针

P

pad　*vt.* 衬垫
pad stitch　*n.* 纳针
palette　*n.* 调色板
palm　*n.* 手掌
pants　*n.* 裤子
pattern cutter　样板师
pierce　*vt.* 刺穿
pin　*n.* 别针
pivot　*vi.* 枢轴
plaid　*n.* 格子图案
plain　*n.* 平纹
pleat　*n.* 褶裥
pocket　*n.* 口袋
point　*n.* 点
pose　*n.* 姿势
predictable　*adj.* 可预测的
prêt-à-porter　成衣
prevail　*vi.* 盛行
princess seam　公主线
print　*n.* 印花

proportion　*n.* 比例
punch　*vt.* 打孔
punk　*n.* 朋克
purity　*n.* 纯度
purple　*n.* 紫色

R

raglan　*n.* 插肩袖
rayon　*n.* 人造丝
retail　*n.* 零售
rhythm　*n.* 节奏
ribbon　*n.* 缎带
right-angle　直角
rivet　*n.* 铆钉
roll line　翻领线
rosette　*n.* 玫瑰花结

S

sailor　*n.* 水手
sample　*n.* 样衣
saturation　*n.* 饱和度
seam　*n.* 接缝
seamstress　*n.* 女裁缝
selvage　*n.* 布边
separates　*n.* 可搭配穿着的女服
set-in　装袖
shave　*vt.* 剃须
shirt collar　衬衫领
shoe　*n.* 鞋
shoulder　*n.* 肩部
shoulder blade　肩胛骨
shoulder neck point (SNP)　颈肩点
showroom　*n.* 陈列室
silhouette　*n.* 轮廓
silk　*n.* 丝绸
size chart　尺寸表

skirt   n. 裙子
slash   vt. 切口
sleazy   adj. 质地薄的
sleeve   n. 袖子
slender   adj. 苗条
sloper   n. 原型
snap fastener   揿纽
spandex   n. 斯潘德克斯弹性纤维
splice   vt. 拼接
stand-up   站领
stitch   n. 缝迹线
strapless gown   无吊带裙袍
straignt grain   直丝缕
stream   n. 潮流
stretch   n. 拉伸
style   n. 风格
subdued   adj. 柔和色
suede   n. 小山羊皮
suit   n. 套装
swatch   n. 样本
swimsuit   n. 泳装

T

tailor tack   线钉
tape measure   卷尺
tattoo   n. 文身
tension   n. 张力
template   n. 模板
textile   n. 纺织品
texture   n. 纹理
theme   n. 主题
thick   adj. 厚的
thin   adj. 薄的
thumb   n. 拇指
tiara   n. 冠状头饰
toe   n. 脚趾
toile   n. 坯布

tone   n. 色调
torso   n. 躯干
touch   n. 手感
trace   vt. 拓印
transparent   adj. 透明的
trend   n. 趋势
trim   n. 装饰
trouser   n. 裤子
trunk show   非公开时装表演
t-shirt   T恤
tuck   n. 箱型褶裥
turquoise   n. 蓝绿色
twist   vt. 加捻
type   n. 类型

U

underarm   adj. 腋下的
undercollar   n. 底领
undergarment   n. 内衣
uniform   n. 制服
upper arm   上臂

V

velvet   n. 丝绒
vertical   adj. 垂直
violet   n. 紫罗兰
vogue   n. 时尚
volume   n. 体积

W

waist   n. 腰部
waistband   n. 腰带
waistline   n. 腰围线
wardrobe   n. 衣橱
warm   adj. 暖色

warp *n.* 经纱	wig *n.* 假发
wear *vi.* 穿着	woven *v.* 梭织
weave *n.* 织物	wrinkle *vi.* 起皱
weft *n.* 纬纱	wristline *n.* 腕围线

References　参考资料

1. http://en.wikipedia.org/wiki/Clothing.
2. http://en.wikipedia.org/wiki/Fashion.
3. https://www.google.com/fashion.
4. http://www.fashiondesignscope.com/?p=117.
5. http://house-of-jo.blogspot.com/2012/09/.
6. http://www.emmaseabrooke.com/.
7. http://www.clothingpatterns10.com/.

说明：本书的资料收集自不同渠道，经作者改编。由于收集资料历时漫长，且最初都是出于教学需要采集，很多资料没有记录相应出处，所以无法在上述参考文献中一一列出，在此深表歉意！

Postscript 后 记

本书力求涵盖尽可能多的服装专业知识和专业词汇，因此精选了服装文化、服装设计、服装发布、服装纸样和服装缝制等各方面的内容。读者既能阅读原汁原味的服装专业文章，又能了解相关专业知识或学到相关专业技能。

感谢庄娜、陈锦、洪玉娟、夏萌、吴佳慧等五位同学为本书配图所做的工作。感谢柳皋隽、高淑琴、田云、郭有庆、王兴国、厉庆荣等为此书所做的工作。